THE BIOLOGY OF MOSSES

THE BIOLOGY OF MOSSES

by D. H. S. RICHARDSON

Professor of Botany
Trinity College, Dublin

BLACKWELL SCIENTIFIC PUBLICATIONS
OXFORD LONDON EDINBURGH
BOSTON MELBOURNE

FOR JOHN, DOROTHY
AND ROXANNE
with love and appreciation

Editorial offices:
Osney Mead, Oxford, OX2 0EL
8 John Street, London, WC1N 2ES
9 Forrest Road, Edinburgh, EH1 2QH
52 Beacon Street, Boston
Massachusetts 02108, USA
214 Berkeley Street, Carlton
Victoria 3053, Australia

First published 1981

British Library
Cataloguing in Publication Data

Richardson, D. H. S.
The biology of mosses.
1. Mosses
I. Title
588'.2 QK533

ISBN 0-632-00782-6

Typeset by Enset Ltd
Midsomer Norton, Avon
Printed and bound in Great Britain by
William Clowes (Beccles) Limited,
Beccles and London

CONTENTS

PREFACE

The last two decades have seen a resurgence of interest in the lower plants and particularly in the lichens. Thus books have been published on the effects of air pollution on lichens, on their ecology and on their biology. In contrast, no book on the biology of mosses is currently in print. Much research has now been completed on the physiology and biology of these plants and this has provided the data base for the present book. Also included are accounts of the value of mosses for pollution monitoring and of the traditional uses of these plants by man. Lighter material has been interwoven with the more academic aspects of the subject to make the book more enjoyable and easier to assimilate for both students and naturalists.

The texts of mosses in English, available to students, are rather limited. The small volume by Dr E.V. Watson, *The Structure and Life of Bryophytes*, deals mainly with structural and morphological aspects while that by Dr N.S. Parihar, *Introduction to the Embryophyta, Vol. 1, Bryophyta*, provides excellent and detailed accounts of a selected number of bryophytes but has very limited information on the physiology, biology or economic importance of the group. Finally, the volume by Dr P. Puri, *Bryophytes: a Broad Perspective*, is now out of print. There is thus a need for an additional volume and it is hoped that the present book will not only fill the gap, but also stimulate further interest in the experimental and applied aspects of this remarkable group of plants.

ACKNOWLEDGEMENTS

The generous help of the following people enabled the writing of this book during a sabbatical leave from Laurentian University, Sudbury, Ontario, Canada.

I would like to thank Professor J.H. Burnett who arranged for me to occupy space and benefit from facilities in the Department of Agricultural Science, Oxford University, which is situated close to the various excellent libraries. The Principal and Fellows of Linacre College, Oxford elected me to the Senior Common Room which promoted my wellbeing in other respects for the duration of my stay. Space, facilities and an Honorary Research Associateship were also kindly arranged for me by Dr J.W. Green, Director of the Western Australian Herbarium, Department of Agriculture, Perth during the latter part of my sabbatical which was spent in Australia.

I am particularly indebted to Dr M.C.F. Proctor who contributed a large proportion of the photographs of mosses used in the book. He, Dr H. Crum, and Dr M.R.D. Seaward made many useful comments on the whole manuscript while Professor D.J. Cove and Professor G.W. Byers made suggestions for improvement of Chapters 6 and 7 respectively. I would also like to thank Dr V. Sarafis who read various parts of the manuscript and directed me to important literature in German. The following people generously contributed original photographs together with advice and information, or copies of photographs from papers which they had published: Dr H. Ando, Dr G. Argent, Dr J.H. Bland, Professor G.W. Byers, Dr S.A. Corbet, Professor D.J. Cove, Professor J.G. Duckett, Dr C. Hébant, Dr C.J. Heusser, Dr P. Hicklenton, Mr D. Hosking, Dr J.L. Gressitt, Dr A. Koponen, Dr B. Lange, Dr R.E. Longton, Dr K.E. Powell, Dr J.D. Shorthouse, Dr L.C. Spiess, Dr I.G. Stone and The Chief Engineer, Irish Electricity Supply Board.

I would also like to thank others who contributed helpful information or reprints including Dr F. S. Chapin, Dr G. Clarke, Professor D.J. Cove, Miss V.A. Hinton, Dr N.D. Jago, Dr W.C. Oechel, Dr M. Owen, Dr M.C.F. Proctor and Dr M.R.D. Seaward. Finally, I would like to express my appreciation to Roxanne who undertook the time consuming task of checking the manuscript and the proofs.

CHAPTER 1
STRUCTURE, HISTORY AND ILLUSTRATION

1 STRUCTURE AND CLASSIFICATION

Mosses are widely recognized even though the individual plants are small and not conspicuous. This is because mosses often form extensive green carpets which bear the characteristic brown capsules in late spring. Spores released from the capsules germinate to give green filaments termed protonemata. The green leafy moss shoots which will bear the capsules arise from the protonemata. There are some 15,000 species of mosses and the variation in the size and shape of both the leafy shoots and capsules is very great indeed. The largest of all mosses is *Dawsonia* from Australasia whose shoots commonly grow to more than 20 cm high but may reach 70 cm when found under ideal conditions in wet shady forests. The aquatic moss *Fontinalis*, common in the Northern Hemisphere, can grow to a similar size. At the other extreme are the earth mosses such as *Ephemerum* which may be less than half a millimetre high.

The stems of the large robust mosses such as *Dawsonia* or the cosmopolitan hairy cap moss, *Polytrichum*, are composed of several different types of tissue. Each is specialized for the conduction of water, food or for assisting with mechanical support. However, the stems of most mosses are not like this but are made up of undifferentiated cells with the exception of a small central strand and a narrow band of thickened cells in the outer region. Even these modifications are absent in many genera. Thus the mechanical and conducting tissues of moss stems do not compare in efficiency or complexity with those found in ferns, fern allies, coniferous plants or flowering plants. In addition, mosses lack true roots and only possess tufts of rhizoids which attach them to their substrates. The leaves of mosses are spirally arranged and usually lance-shaped with pointed tips. They often sheath the stem at the base, so providing spaces where water is held by capillary attraction. Moss leaves are normally one cell thick except in species where there is a thickened border or leaf vein. Often the margin is toothed and the cells from which the leaf is composed may be of distinctive shape or sculpturing which aids in the identification of various species.

There are three major groups of mosses (Fig. 1.1). Firstly, there are the granite mosses (Andreaeaopsida), *c.* 100 species which form widely scattered blackish or dark red-brown patches on rocks in mountainous or arctic regions. The blackness, habitat and small size of the individual plants are characteristic of the group. The capsules are unique among mosses in that they open by a number of longitudinal slits. The second group are the peat mosses (Sphagnopsida) *c.* 350 species, which are abundant in acid mires (bogs) and are immediately recognizable by their distinct features. They form the bright green (or occasionally reddish) clumps on mires or on acid boggy ground. At the apex of the *Sphagnum* plant, the branches are usually tightly packed together to form a moplike head. Below these, there are divergent branches which spread widely from the stem and, slimmer, inconspicuous branches that run down parallel to the stem. All these are clothed in leaves composed of large dead cells surrounded by narrow green (or sometimes red-pigmented) living cells. The dead cells have pores and thickenings and readily become filled with water. Thus the water holding capacity of a *Sphagnum* plant is up to 20 times its dry weight and, this characteristic has been made use of by man in many ways (see Chapter 12). The lids of the capsules of peat mosses are thrown off and the spores violently dispersed all at one time. The third and largest group of mosses are the true mosses (Bryopsida) *c.* 14,000 species. The members of this group vary enormously in size and shape but there are two basic growth habits; tuft forming (acrocarpous) and carpet forming (pleurocarpous). In the former, buds formed on the extending green filaments, which develop from a germinating spore, grow upwards to form a compact tuft. The individual shoots are unbranched to sparingly branched with groups of rhizoids at the base. These sometimes run along the surface of the shoot assisting capillary conduction of water up the tuft. Such mosses are frequently epiphytes on rocks or on the crowns of trees. In contrast, the shoots of the carpet forming mosses branch frequently and tend to be horizontally orientated. A series of loosely overlapping plants make up the moss carpets which are a feature of many forest floors.

The capsules that develop on the leafy shoots of true mosses are borne on short stalks (setae) and are protected by a cap of tissue called the calyptra. This falls off to reveal the fully grown capsule which usually has a lid. This in turn is shed and normally exposes a series of

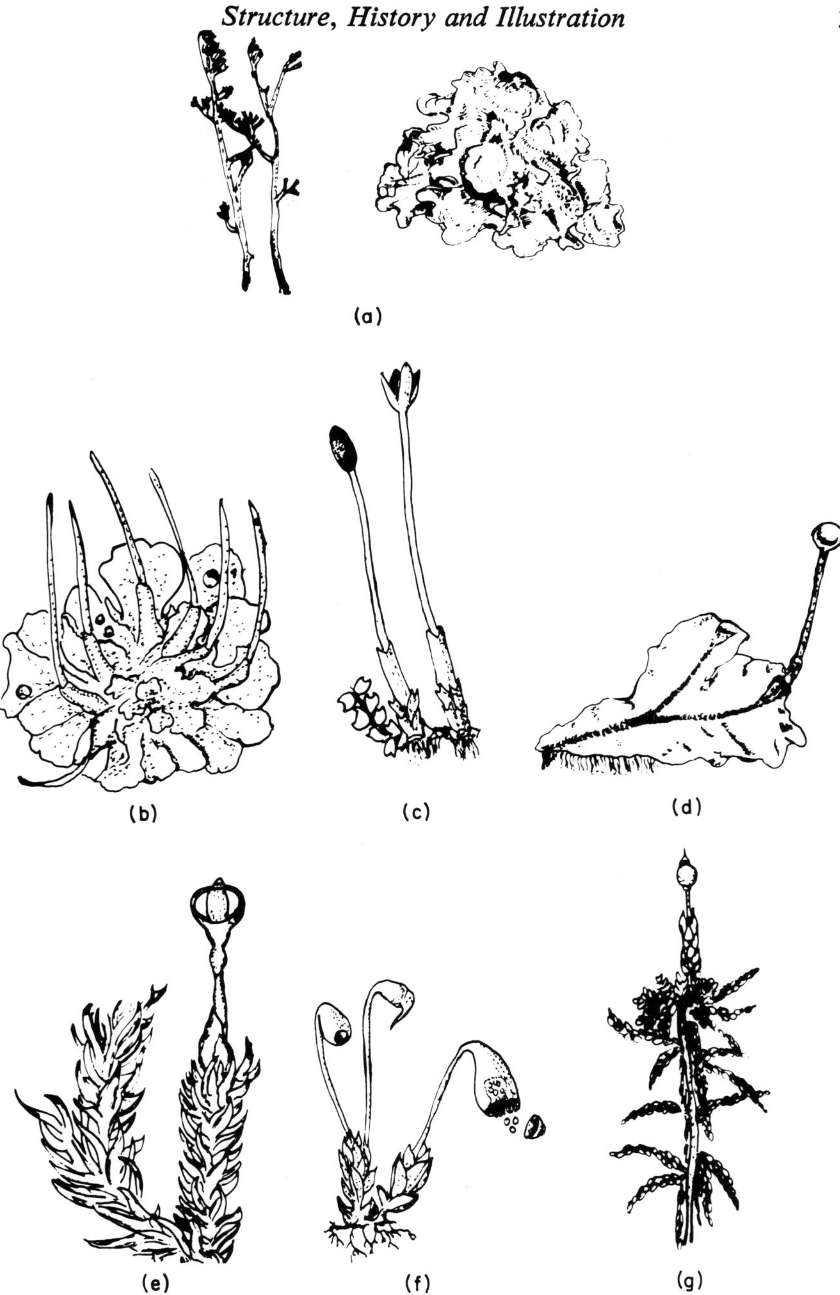

Fig. 1.1 Line drawings to illustrate the macroscopic differences between lichens (a), hornworts (b), leafy liverworts (c), thalloid liverworts (d), and mosses. The main groups of mosses are granite mosses (e), true mosses (f), peat mosses (g) (drawn by R. West).

hygroscopic teeth collectively termed the peristome which help to ensure gradual dispersal of the spores under environmental conditions favourable for their further development (see Chapter 5). The number of teeth is a multiple of four and most frequently 16 or 32. These may be differentiated into larger outer teeth (outer peristome) and more delicate inner teeth (inner peristome). The spores in the capsule are contained in a spore sac which surrounds a central region, the columella, whose function is conductive and mechanical in most mosses but, in a few (the Polytrichales) is expanded at its upper end into a drum like epiphram which is part of the dispersal mechanism.

The 14,000 species of true mosses are classified principally on the characteristics of the capsule and particularly on the features of the peristome. Three subclasses are recognized (the Polytrichideae, the Buxbaumiideae and the Eubryideae) containing, even in a small area of Europe, a total of 19 orders and 50 families and some 200 genera (Smith 1978). In the Polytrichideae, the peristome teeth are robust and do not have transverse thickenings while in the Eubryideae, the teeth are thinner and have transverse bars. The Buxbaumiideae are a small group of true mosses with features intermediate between those of the other two subclasses. (See Appendix 1.)

The thirteen orders in the Eubryideae may be further separated into a group which generally have one set of peristome teeth (Archidiales, Dicranales, Fissidentales, Pottiales, Grimmiales, Encalyptales and Schistostegales) and another group where the peristome is normally double consisting of larger outer and thinner inner teeth (Funariales, Bryales, Orthotrichales, Isobryales, Hookeriales, Thuidiales and Hypnobryales). Each order is divided into families and genera on the characters of both the capsule and the leafy shoot. At the species level, diagnostic criteria include the leaf shape, its vein and the types of cells from which the whole or part of the leaf is constructed. A summary of the main features of the different orders of mosses is displayed in Appendix 1. The genera listed as belonging to each order are those whose biology is discussed elsewhere in this book so that the reader may later refer to this table and appreciate which mosses are closely and which are more distantly related.

Liverworts are often mistaken for mosses. There are two morphological groups—the leafy liverworts and the thalloid liverworts. The latter usually consist of a branching, flat, elongated plate of tissue which grows closely adpressed to damp soil. The leafy liverworts bear much more resemblance to mosses but close examination reveals that the leaves are seldom single-pointed (frequently they are two lobed, and have no midrib) and are normally in two or three rows rather than being spirally arranged on the stem as in mosses. In addition the capsules of liverworts release their spores by splitting into four valves and possess specially thickened cells, the elators, which facilitate spore dispersal. However, the life history of both liverworts and mosses is similar so that they are placed in the same phylum the Bryophyta along with another small rare group the hornworts. They look like thalloid liverworts but have long pointed capsules which have no stalk and split into two valves at the apex (Fig. 1.1). Lichens are another group of plants which are often confused with mosses and are often discussed with them. They are, however, quite distinct in being a symbiotic association between a fungus and an alga that forms a composite plant which does not have leaves. In addition lichens are usually greyish rather than being bright green and they produce fungal fruit bodies not capsules (Richardson 1975).

2 HISTORY OF THE STUDY OF MOSSES

The Greek, Theophrastus, who lived *c.* 300 B.C. does not mention mosses (Hort 1916). Nor are they shown in the illustrations of the writings of Dioscorides who was a physician to the Roman legions and died about 77 A.D. (Dioscorides 1763–1773). He travelled widely with the legions during the time of Vespasian and Nero and described and illustrated the plants known in his day. Mosses also do not seem to have been important in medicine in Saxon times (Cockaine 1864) nor are they illustrated in the earliest herbals (e.g. Anon. 1484). Lichens on the other hand, which were of more value in medicine, were illustrated in such renaissance books. Later, mosses as well as lichens were illustrated in herbals. They were listed under Musci which is Latin for mosses but the group Musci was not well defined in these herbals. For example, in the Niewe Herball (1576) there are several illustrated descriptions that are now known to be one moss (*Polytrichum*), two lichens, one fern ally, one flowering plant and two seaweeds (Dodoens 1576). The illustration of the moss, then called the golden maidenhair moss, is shown in Fig. 1.2 but, according to the herbal 'men use not this moss in medicine'.

The development of botany as a science distinct from pharmacognacy (the study of plants to discover drugs of use in the treatment

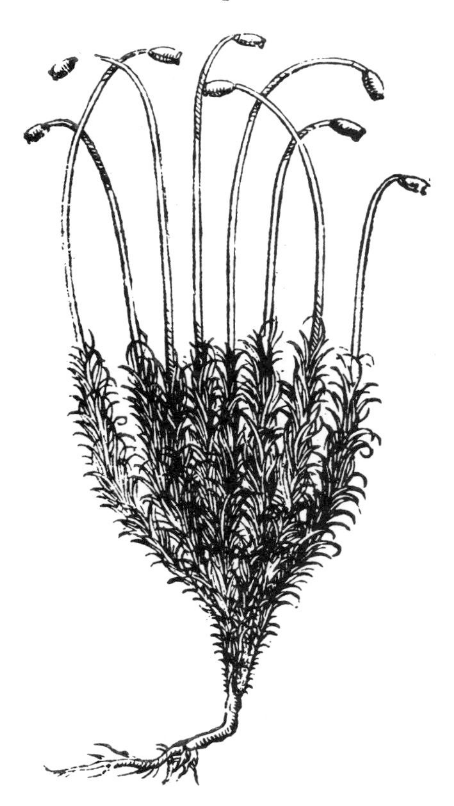

Fig. 1.2 The earliest illustration that could be found of a moss with both leafy shoots and capsules. The moss was then called the golden maidenhair moss and is now known as the hairy cap moss, *Polytrichum* (Dodoens 1576).

of disease) led to extensive collections and detailed descriptions of plants without reference to their medicinal value. Professor Johann Dillen was induced to come to Oxford from Germany and in the tradition of the times latinized his name to Dillenius. He wrote a book entirely devoted to mosses and their allies (Dillen 1741) entitled *Historia Muscorum*. This contains the descriptions of the lichens, club mosses and quillworts (fern allies), liverworts and mosses which were known in his time. Twenty-seven plates were used to illustrate the mosses and familiar genera such as *Bryum*, *Fontinalis*, *Hypnum*, *Mnium*, *Polytrichum* and *Sphagnum* are to be found. A clear improvement, over the early herbals, is evident in the quality of illustration and in the number of features observed (Fig. 1.3). The function of the observed structures intrigued the early botanists and was the subject of lively

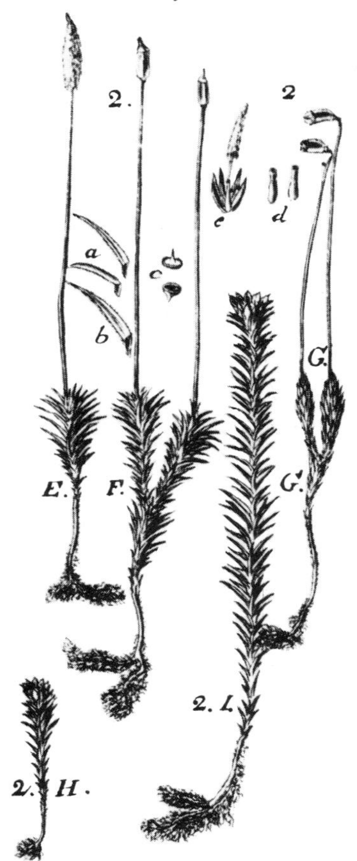

Fig. 1.3 A later illustration of *Polytrichum*, 1741. Note the greater detail of the moss shoots and the capsules with opercula and calyptrae (Dillen 1741).

debate and investigation. They attempted to discover the complete life history of mosses but this proved a difficult task (see next section): indeed algae, liverworts, mosses and ferns became known as the cryptogams which literally means 'the hidden sex' plants.

Towards the middle of the last century, moss study had changed direction once again as the pursuit of natural history and plant collecting by amateurs came into vogue. Thus Stark (1860) in his *Popular History of British Mosses* says that '*Ptilium crista-castrensis* is one of the prizes that botanical tourists should endeavour to carry south (from Scotland), as a remembrance of the land of brown heath and shaggy wood, for its feathery branches suit admirably either for the scientific cabinet or the drawing-room scrapbook'. Slightly earlier, Gardiner in his book *Twenty Lessons on Mosses*, wrote 'were the youthful mind more generally directed to natural objects . . . rather than to matters of frivolous or vicious tendency, it is impossible to say to what extent this might not be conducive to the further advancement and well being of human society'. His book is unusual in that pressed specimens of the mosses described, rather than illustrations, were incorporated in each copy (Gardiner 1848). In his second book on the same subject he concludes that it is pleasant to reflect that 'many of my young readers have thus been led to spend some portion of their leisure hours in a study well calculated to afford them not only a healthy and innocent recreation but greatly to refine and elevate their minds' (Gardiner 1849).

Mosses are easy to collect and may be readily preserved simply by drying which is one reason why extensive collections were made by taxonomists during the late 19th and early 20th centuries from almost every country. The specimens they collected were described, named and incorporated into major herbaria around the world providing an invaluable data base for the preparation of moss floras. The names of many of the collectors are immortalized and familiar, though in abbreviated form, as the cited authority for the moss names used in published works (Rudolph 1969). Europe and North America are bryologically well known but there is still much to be done by taxonomists there (Vitt 1981) while in remote places such as New Guinea, a collector still has little difficulty in finding undescribed species.

During the first half of the present century, studies on the physiology and genetics of flowering plants attracted most attention and mosses were rather neglected. However, in the last two decades many techniques have been adapted for investigations on mosses (Sarafis 1971). Thus great advances have been made in understanding the physiology, genetics and biology of mosses so that bryology is once again regarded as an exciting and popular field of study within botany.

3 DISCOVERY OF THE LIFE HISTORY OF MOSSES

Although the herbals of the renaissance period illustrated a number of mosses, in some cases with capsules (see Fig. 1.2) the life history of these plants was a mystery. Dillen, in his investigations during the early 1700's illustrated both the leaf and the capsule-bearing portions of mosses in great detail (Fig. 1.3), (Dillen 1741) but did not see the sex organs at the tops of the leafy stems, and he interpreted the spore containing capsules as being comparable to spore bearing stamens of flowering plants (Margadant 1973). Dillen considered the starry crowned 'Fowers' of *Polytrichum* to be female.

Pier' Antonio Micheli dissected the 'flowers' of *Polytrichum*, *Mnium* and *Tortula* and within them found antheridia (now known to be male sex organs). He published his results in his book, *Nova Plantarum genera*, in 1729. He also discovered archegonia and paraphyses (now known to be respectively female sex organs and sterile protective hairs) (Micheli 1729). However, Micheli supposed that both the antheridia and archegonia were female organs and that the paraphyses were male organs (Fig. 1.4). He observed these structures in some

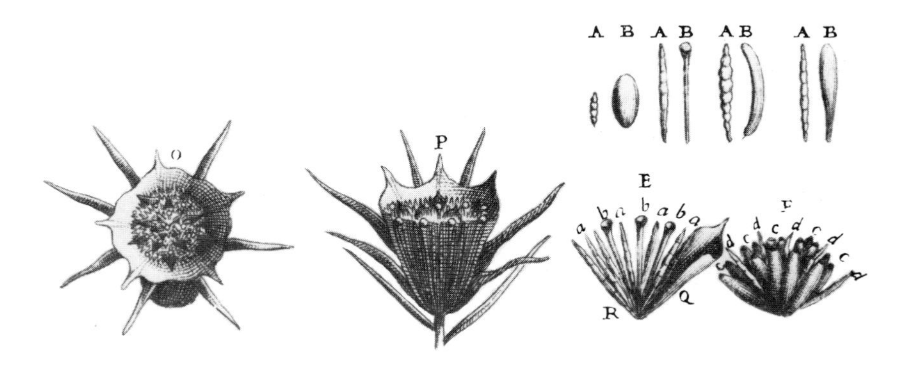

Fig. 1.4 The reproductive structures of mosses as drawn by Michaeli (1729). The male 'flowers' (O,P) contain antheridia and archegonia (B) which he thought were female organs and paraphyses (A) which he believed were male.

30 species. Micheli was a scientist far ahead of his time making important contributions to both bryology (the study of mosses and liverworts) and to mycology (the study of fungi). The description on his memorial in Florence states that he lived 57 years and 22 days, happy though in moderate circumstances, an expert in natural history . . . much loved by all the worthy men of his age on account of his sweetness and modesty (Ainsworth 1965).

The next major achievement in bryology was made some decades later by Johann Hedwig (1782) who confirmed Micheli's observations and also studied the development of capsules. This enabled him to put forward the idea that archegonia were female structures giving rise to capsules from which the spores would again produce the green

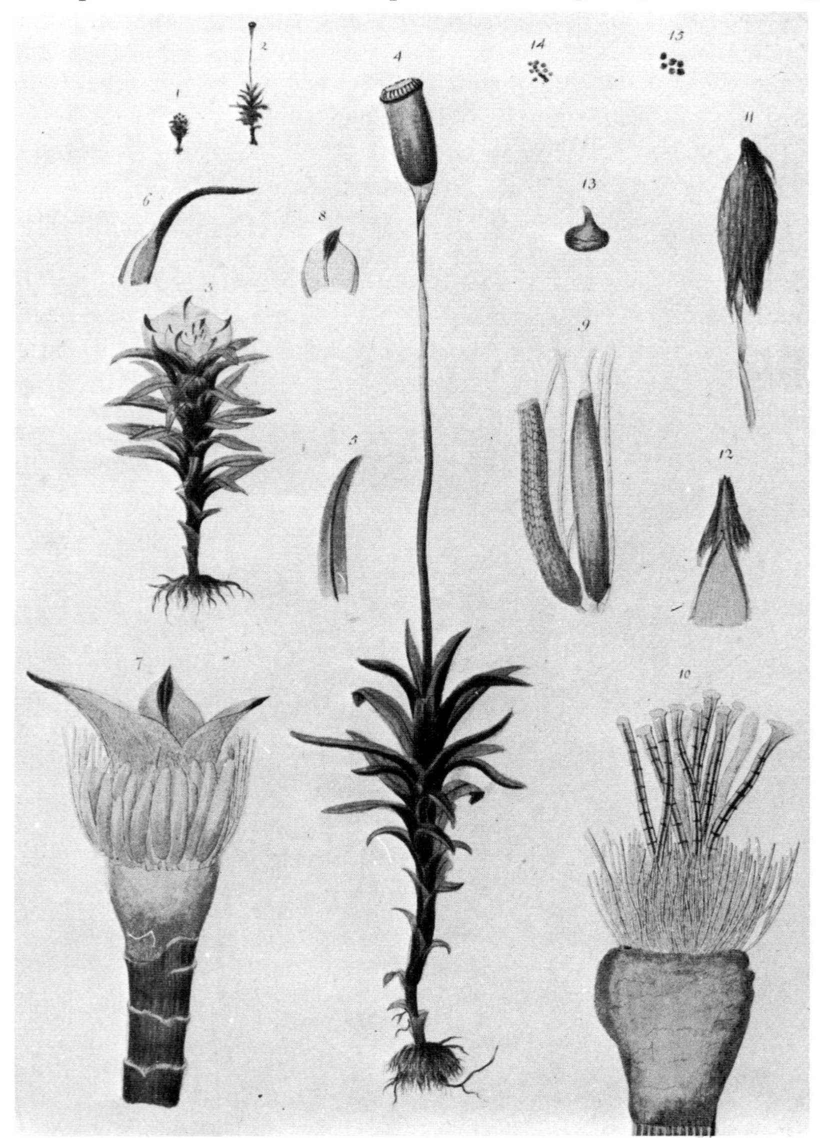

Fig. 1.5 Reproductive structures of *Polytrichum* from a plate by Hedwig (1782) who correctly interpreted antheridia (7,9) as male organs and archegonia (10) as female organs. Note also, (4) the capsule; (11) the calyptra; (13) the operculum and (14) the spores.

leafy plants. He became convinced of the male function of the antheridia in 1776 when he observed the release of sperm cells from antheridia. He must have been a dedicated worker as he learned to draw and engrave rather late in life so that he could illustrate his own work (Margadant 1973). The success of this is evident from the beautiful illustrations of his book published in 1782 (Fig. 1.5). His views were regarded by many in England and Europe to be an error. For example, John Lindley wrote in 1846 that 'there is no doubt that two very different sorts of organs exist among its (moss) species but it does not appear to me that we have sufficient evidence at present to show that antheridia are male organs. So far as they (the antheridia) ar concerned, we have conjecture, nothing more' (Lindley 1846). The matter was finally settled by Wilhelm Hofmeister who, in 1851, described and illustrated the egg cell within the archegonium (Fig. 1.6) and also strongly advocated the idea of the alternation of two generations in

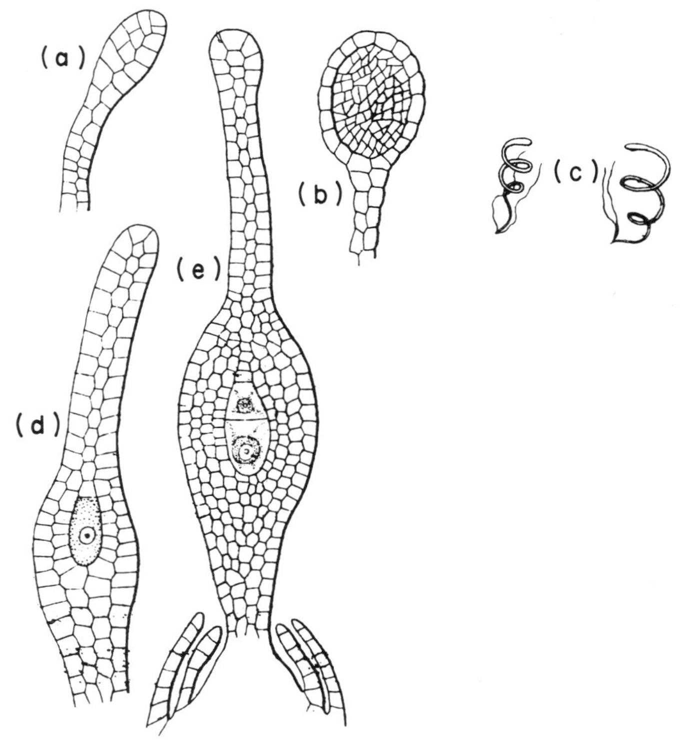

Fig. 1.6 Details of the reproductive organs of *Sphagnum* revealed by cutting sections. Developing antheridia (a, b), sperm cells (c), developing archegonia with contained egg cell (d, e), which enabled the complete life history to be understood (Hofmeister 1862).

the moss life history (Hofmeister 1851, 1862). He conculded that one generation consists of the conspicuous green leafy plant which was termed the gametophyte because at its apex the egg and sperm producing organs were found. The spore-containing capsule which develops from the fertilized egg and which is attached to the parent gametophyte drawing water and some nourishment from it makes up the other generation, termed for obvious reasons the sporophyte. Put in precise terms the gametophyte generation includes: the spore; the green filament (the protonema) first produced on the germination of the spore; the moss plants with stems and leaves (gametophores) together with the rhizoids which develop on the protonema; the sex

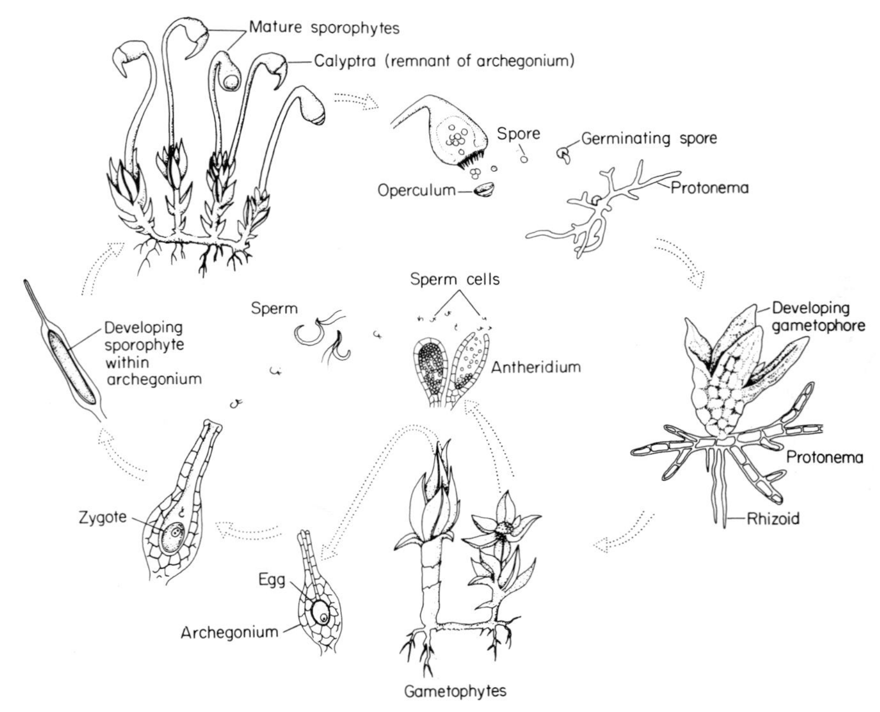

Fig. 1.7 The life cycle of mosses, spores are released from a capsule which opens when a small lid (operculum) is thrown off. The spore germinates to form a branched, filamentous protonema, from which a leafy gametophore develops. Sperm cells, expelled from the mature antheridium, are attracted into the archegonium, where one fuses with the egg cell to produce the zygote. The zygote divides mitotically to form the sporophyte and, at the same time, part of the archegonium divides to form the protective calyptra. The sporophyte consists of a capsule, a stalk, and a foot. Meiosis occurs within the capsule, resulting in the formation of haploid spores (redrawn from Jensen & Salisbury 1972 by C. Blomme).

organs (archegonia and antheridia) on the leafy shoots with their contained eggs and sperms. The sporophyte generation is initiated when the egg is fertilized and continues up to the spore mother cell stage within the capsule. This generation includes: the capsule; its stalk (the seta); the organ by which the stalk is attached to the parent gametophyte (the foot) (Fig. 1.7). It is said that the very detailed studies by Hofmeister which revealed the true nature of the life history of mosses were facilitated by his being very short sighted and thus able to prepare tiny objects for microscopic examination.

4 ORIGIN AND FOSSIL HISTORY

There is a wide gap in structure between the simplest mosses and filamentous green algae from which they are presumed to have arisen. Studies with the electron microscope have indicated that the freshwater green alga *Coleochaete*, its relatives, or possibly *Chara*, are the most likely to have provided the ancestor of mosses and other land plants (Moestrup 1974; Stewart & Mattox 1975). In becoming terrestrial the algal ancestor of mosses evolved a means for protecting the sperm and egg cells and indeed the whole plant from desiccation as well as a mechanism for dispersing spores in the aerial environment.

The origin of the bryophytes has been a topic for speculation for many years (Haskell 1949; Richards 1959; Anderson 1974). The fossil record has not revealed plants which are intermediate between bryophytes and algae. Indeed the earliest land plants which are dated to the Silurian period (*c.* 440–410 million years ago) appear closely related to ferns and fern allies. It is traditionally accepted by palaeobotanists that mosses and liverworts evolved later. However, there is a possibility that at least liverworts may have arisen concurrently with the fern allies. Sheets of cells with a cuticle but without stomata have been discovered in deposits of Silurian–Devonian age (*c.* 440–345 million years old) (Banks 1975; Pratt *et al.* 1978). These could be from bryophytes but sufficiently well-preserved identifiable material has not come to light to make categorical statements. The problem is that the diagnostic capsules of mosses and liverworts exist for only a short portion of the year. For example, the whole development from the start of capsule expansion to shedding of the spores takes place in as little as five weeks in *Mnium hornum* (Proctor 1977). Capsules are extremely rare in fossilized material. This is the case even in fossils of Tertiary (up to 65 million years old) or Quaternary age (*c*. 2 million years old) where moss leaves and shoots are beautifully preserved (Lacey 1969). Also, to be well preserved, delicate plants like mosses have to be embedded in fine sediments in fresh water under anaerobic

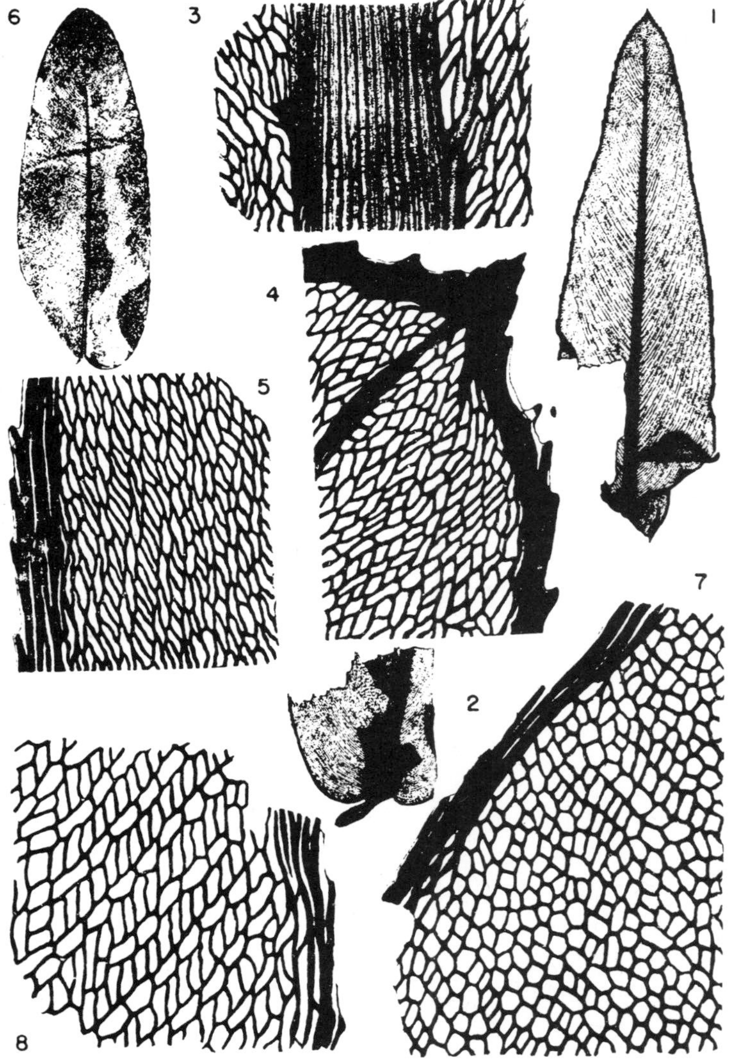

Fig. 1.8 Fossilized moss leaves from the upper Palaeozoic strata in U.S.S.R. (1–6) *Polyssaevia spinulifolia* related to the extant true moss *Mnium* (Bryopsida). (7–8) *Protosphagnum nervatum* with characteristics of a peat moss (Sphagnopsida) (Neuberg 1958).

conditions. The extreme rarity of mosses in the plant remains of the subtropical upper Palaeozoic era is apparently connected with the lack of conditions favourable for their burial (Lacey 1969).

The fossil record for mosses is however more substantial than is generally realized. Well preserved material identified as moss tissue on the basis of leaf and cell wall structure (Fig. 1.8) is widespread both geographically and stratigraphically. Apart from a number of fossils of uncertain affinity, the oldest fossil moss is probably *Muscites polytrichaceus* which has vague similarities to *Polytrichum*. It was found in the upper Carboniferous deposits in France dated about 320 million years old. Three species of *Sphagnum*-like plants have been discovered in rocks of the Permian age (280–225 million years old) while undoubted members of this genus have been found from Tertiary and Quaternary deposits. Of the true mosses (Bryopsida) the oldest also come from the Permian era and are best preserved from various localities in Angardia, U.S.S.R. (Neuberg 1958). Of seven genera found, some specimens have striking resemblances to the extant genus *Mnium* (Fig. 1.8). The fossil mosses of Tertiary and Quaternary times are abundant and can be assigned to living genera. Finally, although liverworts in general appear earlier in the fossil record than mosses, the diversity of the moss flora from Permian rocks in Russia would imply an earlier evolutionary origin, probably in the Devonian period about 350 million years ago.

5 REFERENCES

AINSWORTH G.C. (1965) Historical introduction to mycology. In *The Fungi, vol. 1*, eds Ainsworth G.C. & Sussman A.S., pp. 3–20. Academic Press, N.Y.

ANDERSON L.E. (1974) Bryology: 1947–1972. *Ann. Missouri Bot. Gard.* **61**, 56–85.

ANON. (1484) *Herbarius*, Louvain.

BANKS H.P. (1975) The oldest vascular land plants: a note of caution. *Rev. Palaeobot. Palynol.* **20**, 13–25.

COCKAINE O. (1864) *Leechdoms, Wortcunning and Starcraft in Early England, vol. 1.* Longman, Green, Longmans, Roberts & Green, London. Published under the direction of the Master of the Rolls, Vol. 35a, 1–325.

DILLEN J.J. (Dillenius) (1741) *Historia Muscorum*, Oxonii.

DIOSCORIDES (1763–1773) *Plantarum Dioscorides icones 412, ex manuscript Dioscorides Constantinoplitano Viennae* (Sibthorp's own copy, Botany School, Oxford University).

DODOENS R. (1576) *A Niewe Herball Translated by Henry Lyte.* Gerard Dewes, London.

GARDINER W. (1848) *Twenty Lessons on British Mosses*, 3rd ed. Longman, Brown, Green and Longmans, London.

GARDINER W. (1849) *Twenty Lessons on British Mosses*, second series. Longman, Brown, Green & Longmans, London.

HASKELL G. (1949) Some evolutionary problems concerning the Bryophyta. *Bryologist*, **52**, 49–57.

HEDWIG J. (1782) *Fundamenta Historiae Naturalis Muscorum Frondosorum*. Leipzig.

HOFMEISTER W. (1851) *Vergeichende Untersuchungen der Keimung, Entfalung und fuchtbildung hoherer Kryptogamen*. Leipzig.

HOFMEISTER W. (1862) *On the Germination, Development and Fructification of the Higher Cryptogamia and on the Fructification of the Coniferae*. Translated by F. Currey, Ray Society, London.

HORT A. (1916) *Theophrastus: Enquiry into Plants and Minor Works on Odours and Weather Signs*. William Heinemann, London.

JENSEN W.A. & SALISBURY F.B. (1972) *Botany: An Ecological Approach*. Wadsworth Publishing, Belmont.

LACEY W.S. (1969) Fossil bryophytes. *Biol. Rev*. **44**, 189–205.

LINDLEY J.L. (1846) *The Vegetable Kingdom*. Bradbury & Evans, London.

MARGADANT W.D. (1973) The development of the opinions concerning the sexuality of bryophytes. *Janus* (*Leiden*), **60**, 193–8.

MICHELI P.A. (1729) *Nova Plantarum Genera*. Florence.

MOESTRUP Ø. (1974) Ultrastructure of the scale-covered zoospores of the green alga *Chaetosphaeridium*, a possible ancestor of the higher plants and bryophytes. *Biol. J. Linn. Soc*., **6**, 111–25.

NEUBERG M.F. (1958) *Palaeozoic Mosses from Angarida*. Proc. XX Int. Geol. Congr. Mexico (1956), **7**, 97–106.

PRATT L.M., PHILLIPS T.L. & DENNISON J.M. (1978) Evidence of nonvascular land plants from the early Silurian (Llandoverian) of Virginia, U.S.A. *Rev. Palaeobot. Palynol*., **25**, 121–49.

PROCTOR M.C.F. (1977) Evidence on the carbon nutrition of moss sporophytes from $^{14}CO_2$ uptake and the subsequent movement of labelled assimilate. *J. Bryol*., **9**, 375–86.

RICHARDS P.W. (1959) Bryophyta. In *Vistas in Botany*, Turrill W.B. ed., pp. 387–420. Pergamon Press, London.

RICHARDSON D.H.S. (1975) *The Vanishing Lichens*. David & Charles, Newton Abbot.

RUDOLPH E.D. (1969) Bryology and lichenology. In *A Short History of Botany in the United States*, ed. Ewan J., pp. 89–96,11th Int. Bot. Congress publication. Hafner Publishing, N.Y.

SARAFIS V. (1971) Modern methods in the investigation of bryophyte morphology. *N.Z. J. Bot*., **9**, 725–738.

SMITH A.J.E. (1978) *The Moss Flora of Britain and Ireland*. Cambridge University Press, Cambridge.

STARK (1860) *Popular History of British Mosses*. George Routledge, London.

STEWART K.D. & Mattox K.R. (1975) Comparative cytology, evolution and classification of the green algae with some consideration of the origin of other organisms with chlorophylls *a* and *b*. *Bot. Review*, **41**, 104–35.

VITT D.H. (1981) Sphagnopsida and Bryopsida. In *Taxonomy and Classification of Living Organisms*. McGraw-Hill, N.Y.

CHAPTER 2
WATER RELATIONS

Mosses do not possess roots and except for those mosses with rhizome systems such as *Polytrichum*, *Dawsonia*, *Climacium* etc., have very limited powers of withdrawing water from the substrate. The water required for metabolic processes is therefore derived from water falling on or flowing over, the plants. Additionally, water is required to transfer the sperm cells from antheridia to archegonia. These constraints have resulted in mosses: (a) becoming aquatic; (b) being confined to continually moist habitats; and (c) evolving the ability to be poikilohydric.

The latter means that although the plant dries almost as rapidly as its surrounding environment, it can resume normal metabolic activity on remoistening. This strategy enables mosses to colonize substrates such as rock or bark which are impenetrable to the roots of higher plants. In such situations mosses are free from competition by flowering plants for the available light and nutrients (Proctor 1979a). The response of mosses to drought has been the subject of numerous investigations and several reviews (Richards 1959; Biebl 1962; Walter *et al.* 1970; Crum 1976; Proctor 1979a, 1980, 1981; Longton 1979, 1981).

1 DROUGHT TOLERANCE

The ability of some mosses to survive desiccation is remarkable. Herbarium specimens have been revived after long periods under dark dry conditions: *Anoectangium compactum* retained its viability for 19 years. Other examples include *Dicranoweisia cirrata*, 9 years; *Grimmia elatior* 5 years; *Anomodon longifolius*, *Bryum argenteum* and *Orthotrichum ripestre c.* 2 years; *Grimmia muehlenbeckii* 18 months (Malta 1921). More recent studies in the laboratory have confirmed the drought tolerance of some mosses. *Tortula ruralis* rapidly resumed normal rates of protein synthesis after 8 months of drying while *Rhacomitrium lanuginosum* and *Andreaea rothii* (a granite moss) survived respectively 11 and 13 months desiccation over calcium chloride at 32% relative humidity (Proctor 1981). These tolerant mosses belong to the third strategy mentioned earlier. Mosses growing in deserts such as the Sonoran Desert, U.S.A. have a particular problem as rain is very infrequent. Following an overnight storm it was found that *Bryum capillare*, *Grimmia* spp., *Tortula* spp. and *Weissia controversa* had a water capacity at sunrise (06.00 h) of *c*. 85% of their respective saturation water contents. Eleven of the 12 species investigated contained less than 2 g water/g of plant by 09.00 h, less than 0·5 g water/g by 15.00 h and less than 0·1 g water/g by 17.00 h. It is evident that mosses in deserts desiccate very rapidly and it is estimated that in the above example less than 9 hours was available for net photosynthesis (Alpert 1979). Thus the scarcity of mosses in many deserts is due to the limited time available for net assimilation rather than to damage by drought. Proctor (1981) points out that highly desiccation-tolerant mosses, like 'orthodox seeds' survive best at low water contents. Less tolerant mosses behave like 'recalcitrant seeds' and are more damaged the drier they become. As might be expected, aquatic mosses or those which grow in humid forests are rather easily damaged by drought.

One study on aquatic mosses showed that *Fontinalis novae-angliae* and *F. dalecarlica* died after 55 hours desiccation in the laboratory but when left on the stream bank lived for up to one year (Glime 1971). The fact that this moss survived mild desiccation is perhaps related to it having to cope with fluctuating water levels in streams and also to the way by which it is dispersed. Detached moss stems of this genus when suspended, for example on tree branches above a stream by falling water levels, tend to break up and fragment. On subsequently falling into water the moss pieces can become attached to surfaces against which they lodge through the production of rhizoids: a process which takes about nine weeks. In this way rapid spread through a river system can occur (Glime *et al.* 1979). Indeed when *Fontinalis antipyretica* was introduced into a South African stream, under the impression that it would harbour shrimps and larvae for trout food, it spread so rapidly that it eliminated rock-dwelling species of insects. *Hydropogon fontinaloides* from the Amazon basin has a similar adaptation in that during the dry season it hangs from trees while in the wet season it floats in the water (Mägdefrau 1973). The importance of slow drying is also seen in studies on another aquatic moss from Canada, *Cratoneuron*. It was found to be unable to recommence protein synthesis following rehydration if it was dried quickly over silica gel for one hour. Slow drying to 20% of its fresh weight also resulted in no protein synthesis. However, some synthesis was observed in specimens dried to 33% of their fresh weight while material dried slowly for 5 hours to 66% of its fresh weight showed no detrimental effects as measured by protein synthesis (Bewley 1974).

In tropical forests many mosses normally live under near saturated conditions and may be damaged by as little as 4 hours exposure to relative humidities (RH) of 63%. Examples of such sensitive mosses include terrestrial species (e.g. *Fissidens crassinervius*, *Leucobryum sanctum* and *Semibarbula orientalis*) and epiphytic species (*Calymperes dozyanum* and *C. longifolium*). It is interesting that the latter are found on the wettest side of the trees continually watered by the runoff from above. The one-sided distribution of many bryophytes in tropical forests is quite striking (Johnson & Kokila 1970). The mosses from moist situations in temperate regions are also drought sensitive. The water content of *Hookeria lucens*, common on shaded stream margins in England, may not fall below 100% even in summer. When desiccated in the laboratory at 54% RH, 20°C, severe damage is induced by more than a few days drying (Dilks & Proctor 1974, 1979).

Finally, mosses vary in their desiccation tolerance at different times of year. Generally they are most sensitive during autumn and early winter with their tolerance increasing through spring to a maximum in early summer. The fall in tolerance at the end of the summer, a feature also observed in lichens (Richardson & Nieboer 1980), coincides in mosses with the resumption of growth (Dilks & Proctor 1976a). The increase in tolerance before the onset of the summer dry period is termed acclimation.

2 DESICCATION AND REHYDRATION

As a moss dries, the water content falls and the water potential, which is measured in bars, becomes more negative. At low (negative) water potentials the cytoplasm within the cells becomes viscous and gel-like.

A moss with a water content about 20% of its dry weight would have a water potential of about −200 to − 400 bars and this might be expected for mosses growing in a shaded wood during summer. When the water content falls to 10% dry weight, the water potential is around − 1000 bars and could be found in mosses during dry but not extremely hot conditions in the open in summer (Dilks & Proctor

1979). Drought-tolerant mosses can survive these conditions and resume normal metabolic activity when remoistened. The differences between mosses and flowering plants is put in perspective by the fact that net photosynthesis in the latter decreases as soon as water potentials drop below −1 to −3 bars. Indeed in most flowering plants photosynthesis may reach zero or even become negative (i.e. respiration is exceeding fixation) when potentials drop below − 12 to − 15 bars. Thus, to survive, flowering plants and ferns must control their water contents so that they fall within the limits set by these water potential values (Busby & Whitfield 1978).

Mosses from the circumboreal woodlands of northern Canada, such as *Pleurozium schreberi*, *Hylocomium splendens* and *Tomenthypnum nitens*, carry on the processes of photosynthesis and respiration until the water potential falls to − 55 to − 100 bars (Busby & Whitfield 1978). Species from various habitats in the U.K. have been investigated using techniques in which samples are incubated in glycerol solutions of known osmotic potential. Net photosynthesis was found to be positive down to about −150 bars in *Camptothecium lutescens*, a tolerant species but only to 80 bars in the sensitive species *Hookeria lucens* (Fig. 2.1) (Dilks & Proctor 1979). Dilks and Proctor (1979) have prepared a schematic representation of the relationship between water content and water potential in drought tolerant mosses (Fig. 2.2). The break at about − 200 bars correlates with the limit of metabolic (particularly respiratory) activity and it is suggested that this is the point at which a free liquid phase occurs in the cytoplasm.

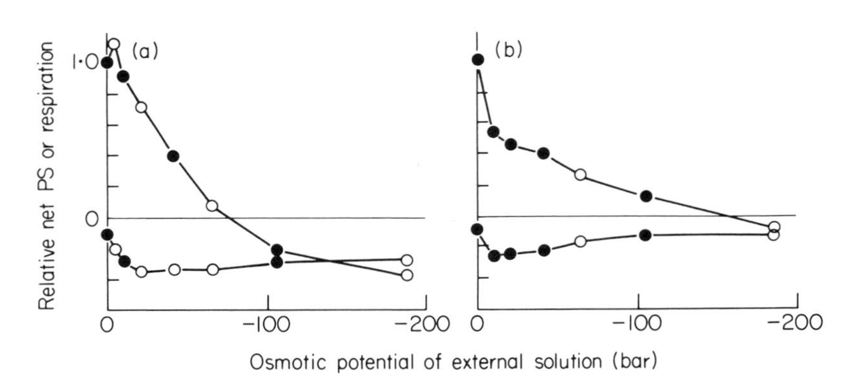

Fig. 2.1 Relation of net photosynthesis and respiration to osmotic stress (glycerol) for (a) *Hookeria lucens* and (b) *Camptothecium lutescens*. ○ single measurement; ● means of pairs of measurement (Dilks & Proctor 1979).

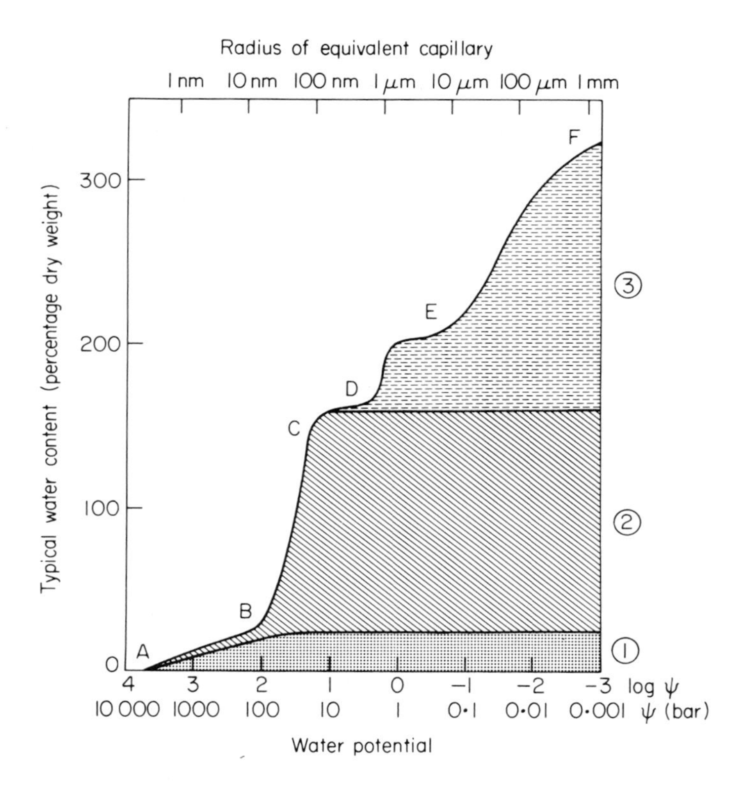

Fig. 2.2 Schematic relation of water content to water potential for a poikilohydrous moss. From A to B most water is closely associated with surfaces and is relatively immobile. At B a free liquid phase becomes important and from B to C increase of water content is mainly within cells and is reflected by the changing potential of the cell solution. C is the point of incipient plasmolysis; from C to D the water content increases only slightly with the build-up of wall pressure and the first appearance of external capillary water associated with fine surface roughness and striation of the cell wall. From D to F the form of the curve will vary from species to species depending on the detailed architecture of the shoot (Dilks & Proctor 1979).

In basal cells of moss leaves the protoplasm condenses at the ends of the cells as the moss dries leaving the central parts empty. At the same time the chloroplasts become round and compact and vesicles disappear (Tucker *et al.* 1975). The cells remain in this condition until remoistened. The following sequence of events occurs when a drought tolerant moss is rewetted. There is initially a release of CO_2 which is thought to be due to displacement of absorbed gas. Using Nomarsky optics to observe moss cells during rehydration, pockets of gas are seen to form immediately after rehydration at the distal and proximal ends of leaf cells of the desiccation tolerant moss *Tortula ruralis*. However, these shrink and disappear within 10 to 30 seconds as the cytoplasm expands to fill the whole cell (Tucker *et al.* 1975). After this initial release of gas there is a period of enhanced respiration lasting from one to a few hours. During this time the cell organelles regain their normal appearance (Noailles 1978). The more rapid the desiccation rate, the longer and more pronounced the period of enhanced respiration (Krochko *et al.* 1979). During this time the process of photosynthesis commences and accelerates. After 4 to 24 hours, normal rates of respiration and photosynthesis are observed (Fig. 2.3) (Peterson & Mayo 1975; Dilks & Proctor 1976b; Proctor 1981). This pattern is typical not only of mosses but also of another group of poikilohydrous plants, the lichens (Bewley 1979; Richardson & Nieboer 1980).

During remoistening, mosses leak photosynthate and this is most rapid during the first two minutes following addition of water, thereafter it slows down. For example, after desiccation for two days at 50% RH, samples of *Plagiothecium undulatum* leaked 250 times as much as controls maintained moist. Less severe drying at 97% R.H., 15°C, still resulted in leakage but it was only 2% of that under the more severe regime though the water loss from the shoots was nearly 69%. It is remarkable that in spite of this leakage, the plants from both treatments remained alive and capable of photosynthesizing (Gupta 1977). The leakage occurs because membranes of the cells and cell organelles are damaged and thus made permeable as mosses dry out. However studies on liverworts have shown that once the membranes have been repaired, following remoistening, some absorption of the lost photosynthate may occur. In lichens loss of photosynthate, phosphate and cations, particularly K^+ has been observed on rewetting dried thalli. This loss is probably inevitable for all poikilohydrous plants though it is generally less in drought tolerant species. The same phenomenon of leakage is also observed when dried seeds are re-moistened. Indeed such leakage is routinely measured in batches of pea seeds as a test of viability since leakage correlates inversely with field emergence (Mathews 1977). The period of enhanced respiration observed in poikilohydrous plants and seeds following remoistening enables repair of cell membranes etc. and curtails cell leakage. However, considerable depletion of cell reserves may occur during this period (Farrar &

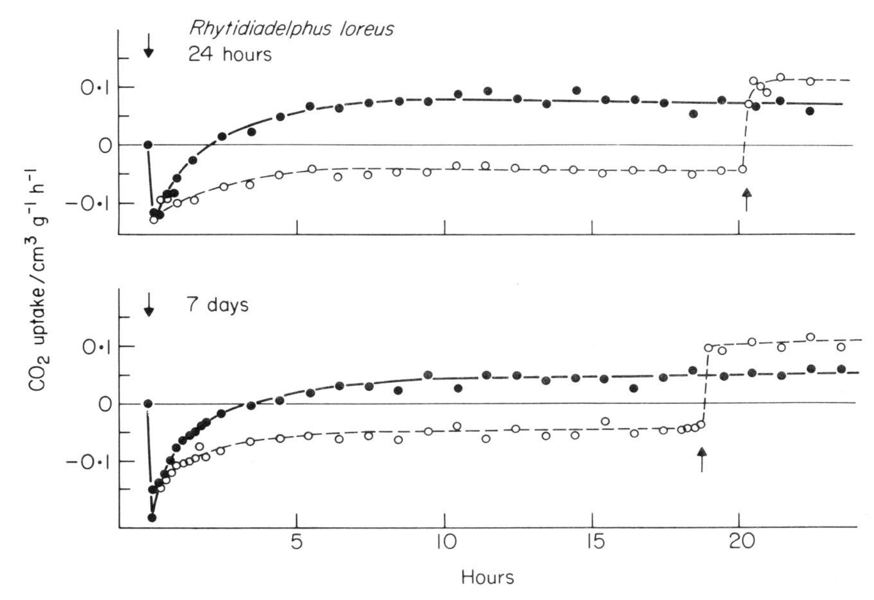

Fig. 2.3 Gas exchange in ***Rhytidiadelphus loreus*** in the light and dark following remoistening after dry periods of 24 hours and 7 days. Arrows indicate when the material was moistened and when the sample measuring dark respiration was exposed to light. Note. Following 24 hours drying, the moss took 2 hours to reach compensation and photosynthesis was steady after 4 hours. Following 7 days drying the moss took twice as long to return to normal and respiration was initially greater (Dilks & Proctor 1976).

Smith 1976; Richardson & Nieboer 1980). In drought tolerant mosses, messenger RNA is apparently conserved during drying and polyribosomes are reformed on rehydration without the need for new RNA synthesis. Thus protein synthesis can begin immediately so that repair can rapidly begin (Dhindsa & Bewley 1978).

3 MECHANISMS OF DAMAGE

It has been suggested that the loss of viability on drying can be due to several causes. Effects have been observed on enzymes such as NADP-linked glyceraldehyde phosphate dehydrogenase (Stewart & Lee 1972) and on the capacity to synthesize proteins (Bewley 1974, 1979) Clearly the loss of photosynthate and essential nutrients during remoistening is an additional problem (Gupta 1977). While the primary cause of loss of viability is open to question, some interesting studies have been completed comparing fast and slow drying on the ultrastructure of two moss species, one drought sensitive, the other tolerant. The species respectively were *Cratoneuron filicinum* and *Tortula ruralis*. Fast drying of *Tortula ruralis* for 24 hours over silica gel resulted in the chloroplasts and mitochondria swelling and losing their integrity when the plant was rewetted. However, within 24 hours these organelles had regained their normal appearance. The chloroplasts and mitochondria of *Cratoneuron filicinum* showed a similar damage on rewetting after fast drying but 25 hours later the cells showed extreme degradation and loss of cell contents. Drying the sensitive species more slowly (for 12 hours over RH of 75%) resulted in only 20% of the cells exhibiting swollen organelles on rewetting and after 24 hour recovery there were approximately equal numbers of degraded cells and cells with normal looking nuclei and mitochondria (Krochko *et al.* 1978). The fact that *Tortula ruralis* can survive swelling of the cell components whereas the sensitive species cannot, suggests that sensitive and tolerant species are separated by the stability, availability and capacity for synthesis of the enzymes needed for repair (Dhindsa & Bewley 1978).

Adenosine triphosphate (ATP) is a compound in which energy is stored within the cell for use in various metabolic processes. It seems more important for dried mosses to retain the ability to synthesize ATP than to conserve normal cellular levels of this substance. In the desiccation-tolerant moss *Tortula ruralis*, control levels of ATP are regained within 30 min even after rapid drying. In the sensitive species *Cratoneuron filicinum*, ATP levels steadily declined following rewetting of fast dried specimens. However slowly dried material was able to resume ATP synthesis and regain normal levels within 2 hours (Krochko *et al.* 1979). It is clear, therefore, that although mosses are poikilohydric, it is an advantage for many species to lose water as slowly as possible. The mechanisms by which mosses retard water loss are summarized in the following section.

4 RESISTANCE TO WATER LOSS

Mosses may, in general, be separated into two groups. Endohydric mosses possess well developed internal water conducting tissue which transfer water reserves from the base of the moss cushion to the actively photosynthesizing leaves at the apex. In the second group, the ectohydric mosses, water is readily absorbed by the stem and leaf surfaces but there is little internal conducting tissue. Often external capillary conducting systems exist in these mosses by which water is distributed to different parts of the plant. Endohydric mosses include most tuft forming true mosses (acrocarpous Bryopsida; i.e. those mosses which form erect tufts that develop archegonia and subsequently capsules at the apex of the main or principal branches) for example the Funariales and Polytrichales. Ectohydric mosses include most carpet forming true mosses (pleurocarpous Bryopsida, i.e. those which have a tendency to dorsiventral orientation and whose archegonia and subsequently capsules form on short lateral branches); some acrocarpous mosses often from very dry habitats, e.g. Grimmiales; the granite mosses (Andreaeopsida) and the peat mosses (Sphagnopsida). The latter may be termed ectohydric hygrophytes since they are confined to living within a few decimetres of a free water surface (Proctor 1979a).

4.1. Endohydric mosses

Retardation of water loss is achieved by the cushion form of many endohydric mosses. This results in an overall surface that is rough so that with moderate wind speeds there is a tendency for a thick layer of still air to form over the colony. Water lost by evaporation must diffuse through this boundary layer and the thicker it is the slower is the evaporation rate. Evaporation from individual leaves of a moss plant is hence much slower when the shoot is part of a tuft than would be the case if it were held erect by itself in wind of the same speed. The large endohydrous mosses like *Polytrichum* and *Dawsonia* have additional adaptations. Firstly, the leaves of some species have marginal wings which fold over the photosynthetic lamellae in the centre (Sarafis 1971) and secondly, the whole leaf can flex part way along its length. These hinge movements are due to the swelling of the lamellae (Zanten 1974) and result in the leaf being reflexed under moist conditions but on drying, the leaves straighten and tend to hug the main axis thus retarding water loss. The tips of the lamellae on these leaves also have a waxy covering but its presence is thought not to be related to water loss. The function of the wax is to repel liquid water which, if it filled the spaces between the lamellae, would interfere with gas exchange (Proctor 1979a, b). However, in some other endohydric mosses, particularly those with a glaucous bloom, the waxy layers may well also retard water loss from the leaves.

4.2 Ectohydric mosses

As mentioned earlier, such mosses absorb water rapidly both into the cells and into the moss carpet by capillary systems of which there are three categories (Buch 1945, 1947): 1 inter-organ capillary systems: e.g. spaces within sheathing leaves or leaf bases or between a tomentum of rhizoids; 2 epi-organ capillary systems: e.g. spaces between papillae, ridges or folds in leaf or stem surfaces; 3 intra-organ capillary systems: e.g. non-photosynthetic cells in *Sphagnum* leaves and the

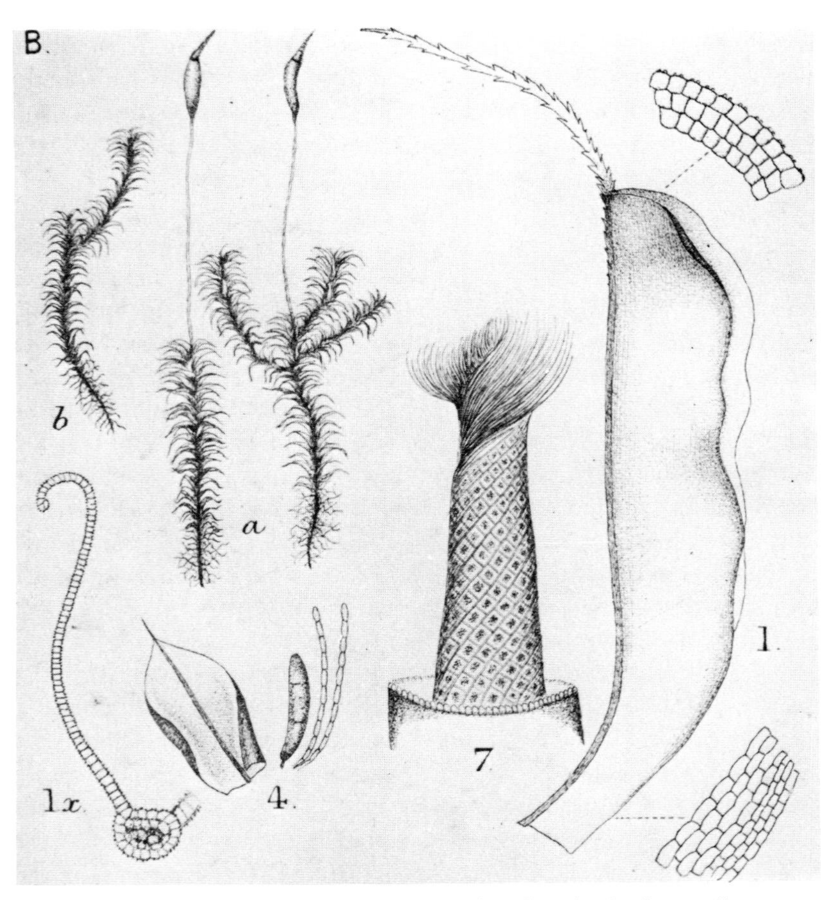

Fig. 2.4 A drawing of *Tortula ruralis*, a plant of old roofs and calcarious wall tops. The leaves (1) have long hair points. (a, b) shoots with and without capsules, (4) antheridia and paraphyses (7) details of peristome showing long twisted teeth (Braithwaite 1887).

porose cells of leaf bases in a number of moss families. Storage in these systems is not great, much more being held in the larger capillary spaces between the moss shoots within the moss carpet (Proctor 1979a). However, it is critical that there be adequate distribution of the water reserves to maintain the photosynthesizing leaves at water levels above the water content at which net photosynthesis ceases. If there is a capillary system which results in a slightly extended period of photosynthesis, it may make the difference between survival and death. Absorption and distribution of small amounts of dew to enable brief periods of photosynthesis, in early morning is essential for many mosses in dry habitats. Certainly studies have shown that absorption of dew is critical for survival of desert lichens and mosses (Lange 1969; Kappen *et al.* 1979). Recent work in Finland on e.g. *Dicranum fuscescens* shows that dew is also very important as a source of water to enable photosynthesis in mosses from temperate forests (Kellomaki *et al.* 1977). Retardation of water loss from ectohydric mosses is enhanced by the tendency to form carpets or tufts so increasing the thickness of the boundary layer of still air over the plants. In addition many species possess leaves with hair points (Fig. 2.4). These reflect a proportion of the incoming solar radiation so that less energy is absorbed and less water evaporation occurs. Also, water falling on the plants accumulates on the upper parts of the leaves rather than on the hair points so that there is an increased diffusion path length for water vapour to reach the boundary layer and hence the rate of water loss is again reduced (Proctor 1979a). Experiments on water loss by plants with and without (surgically removed) hair points shows that the hair tips reduce water loss by about 35% in *Grimmia pulvinata* and *Tortula intermedia* (Proctor 1980).

5 REFERENCES

ALPERT P. (1979) Desiccation of desert mosses following a summer rainstorm. *Bryologist*, **82,** 65–71.

BEWLEY J.D. (1974) Protein synthesis and polyribosome stability upon desiccation of the aquatic moss *Hygrohypnum luridum*. *Can. J. Bot.* **52,** 423–7.

BEWLEY J.D. (1979) Physiological aspects of desiccation tolerance. *Annu. Rev. Plant Physiol.* **30,** 195–38.

BIEBL R. (1962) Protoplasmatische Okologie der Pflanzen wasser und temperatur. *Protoplasmatologia*, **12,** 1–259.

BRAITHWAITE R. (1887) *The British Mosses*, Vols 1, 2, 3, Reeve, London.

BUCH H. (1945, 1947) Uber die Wasser- und Mineralstoffversongung der Moose. *Commentat. biol. Soc. Sci. Fenn.* **9(16),** 1–44; **9(20),** 1–61.

BUSBY J.R. & WHITFIELD D.W.A. (1978) Water potential, water content and net assimilation of some boreal forest mosses. *Can. J. Bot.* **56,** 1551–8.

CRUM H. (1976) Mosses of the Great Lakes Forest. Revised ed. *Univ. Mich. Herb.* 1–404.

DHINDSA R.S. & BEWLEY J.D. (1978) Messenger RNA is conserved during drying of the drought-tolerant moss *Tortula ruralis. Proc. Natl. Acad. Sci. U.S.A.* **75,** 842–6.

DILKS T.J.K. & PROCTOR M.C.F. (1974) The pattern of recovery of bryophytes after desiccation. *J. Bryol.* **8,** 97–115.

DILKS T.J.K. & PROCTOR M.C.F. (1976a) Seasonal variation in desiccation tolerance in some British bryophytes. *J. Bryol.* **9,** 239–47.

DILKS T.J.K. & Proctor M.C.F. (1976b) Effects of intermittent desiccation on bryophytes. *J. Bryol.* **9,** 249–64.

DILKS T.J.K. & PROCTOR M.C.F. (1979) Photosynthesis, respiration and water content in bryophytes. *New Phytol.* **82,** 97–114.

FARRAR J.F. & SMITH D.C. (1976) Ecological physiology of the lichen *Hypogymnia physodes.* III. The importance of the rewetting phase. *New Phytol.* **77,** 93–103.

GLIME J.M. (1971) Response of two species of *Fontinalis* to field isolation from stream water. *Bryologist,* **74,** 383–6.

GLIME J.M., NISSILA P.C., TRYNOSKI S.E. & FORNWALL M.D. (1979) A model for attachment of aquatic mosses. *J. Bryol.* **10,** 313–20.

GUPTA R.K. (1977) A study of photosynthesis and leakage of solutes in relation to the desiccation effects in bryophytes. *Can. J. Bot.* **55,** 1186–94.

JOHNSON A. & KOKILA P. (1970) The resistance to desiccation of ten species of tropical mosses. *Bryologist,* **73,** 682–6.

KAPPEN L., LANGE O.L., SCHULZE E.D., EVENARI M. & BUSCHBOM U. (1979) Ecophysiological investigations on lichens of the Negev desert. 6. Annual course of the photosynthetic production of *Ramalina maciformis* (Del.) Bory. *Flora (Jena),* **168,** 85–108.

KELLOMAKI S., HARI P. & KOPONEN T. (1977) Ecology of photosynthesis in *Dicranum* and its taxonomic significance. *Bryophytorum Bibliotheca,* **13,** 485–507.

KROCHKO J.E., BEWLEY J.D. & PACEY J. (1978) The effects of rapid and very slow speeds of drying on the ultrastructure and metabolism of the desiccation-sensitive moss *Cratoneuron filicinum* (Hedw.) Spruce. *J. Exp. Bot.* **29,** 905–17.

KROCHKO J.E., WINNER W.E. & BEWLEY D.J. (1979) Respiration in relation to adenosine triphosphate content during desiccation and rehydration of a desiccation-tolerant and a desiccation-intolerant moss. *Plant Physiol. (Bethesda),* **64,** 13–17.

LANGE O.L. (1969) CO_2 Gaswechsel von Moosen nach Wasserdampfaufnahme aus dem Luftraum. *Planta (Berl.),* **89,** 90–4.

LONGTON R.E. (1979) Climatic adaptation of bryophytes in relation to systematics. In *Bryophyte Systematics,* eds Clarke G.C.S. & Duckett J.G. Systematics Association special volume No. 14, pp. 511–31. Academic Press, London.

LONGTON R.E. (1981) Physiological ecology of mosses. In *Mosses of North America,* eds Taylor R.J. and Leviton A.E. Pacific Division, American Academy for Advancement of Science, Washington.

MÄGDEFRAU K. (1973) *Hydropogon fontinaloides* (Hook.) Brid., ein periodisch hydroaerophytisches Laubmoos des Orinocos und Amazonas. *Herzogia,* **3,** 141–9.

MALTA N. (1921) Versuche über die Widerstandsfähigkeit der Moose gegen Austrocknung. *Acta Univ. Latv.* **1,** 125–9.

MATHEWS S. (1977) Field emergence and seedling establishment. In *The Physiology of the Garden Pea,* eds Sutcliffe J.F. & Pate J.S., pp. 83–118. Academic Press, London.

NOAILLES M. (1978) Etude ultrastructurale de la récupération hydrique après une période de sécheresse chez une hypnobryale: *Pleurozium schreberi* (Willd.) Mitt. *Ann. Sci. Nat. Ser.* **12,** 19, 249–65.

PROCTOR M.C.F. (1979a) Structure and eco-physiological adaptations in bryophytes. In *Bryophyte Systematics*, eds Clarke G.C.S. & Duckett J.G. Systematics Association special volume No. 14, pp. 479–509. Academic Press, London.

PROCTOR M.C.F. (1979b) Surface wax on the leaves of some mosses. *J. Bryol.* **10,** 531–8.

PROCTOR M.C.F. (1980) Diffusion resistance in bryophytes. In Plants and their atmospheric environment, eds Ford E.D. & Grace J. *Symp. Brit. Ecol. Soc.* 219–229.

PROCTOR M.C.F. (1981) Physiological ecology of bryophytes. In *Advances in bryology*, ed Schultze–Motel W. J. Cramer, Germany.

RICHARDS P.W. (1959) Bryophyta. In *Vistas in Botany*, ed. Turrill W.B., pp. 387–420. Pergamon Press, London.

RICHARDSON D.H.S. & NIEBOER E. (1980) Surface binding and accumulation of metals in lichens. In *Cellular Interactions in Symbiotic and Parasitic Associations*, eds Cook C.B., Pappas P.W. & Rudolph E.D., pp. 75–94. 5th Ann. Colloq. Coll. Biol. Sci. Ohio State Univ. Press.

SARAFIS V. (1971) A biological account of *Polytrichum commune. N.Z. J. Bot.* **9,** 711–24.

STEWART G.R. & LEE J.R. (1972) Desiccation injury in mosses. II. The effects of moisture stress on enzyme levels. *New Phytol.* **71,** 461–6.

TUCKER E.B., COSTERTON J.W. & BEWLEY J.D. (1975) The ultrastructure of the moss *Tortula ruralis* on recovery from desiccation. *Can. J. Bot.* **53,** 94–101.

WALTER H., KREEB K., ZIEGLER H. & VIEWEG G.H. (1970) Die Hydration und Hydratur des Protoplasmas der Pflanzen und ihre oko-physiologische Bedeutung. *Protoplasmatologia*, **2,** 1–136.

ZANTEN B.O. van (1974) The hygroscopic movement of the leaves of *Dawsonia* and some other Polytrichaceae. *Bull. Soc. Bot. Fr., Colloque Bryol.* **121,** 63–6.

CHAPTER 3
PHOTOSYNTHESIS, TEMPERATURE AND NUTRIENTS

Light, temperature and nutrient supply are factors which, in addition to water (see Chapter 2), may cause stress and reduce the photosynthetic performance of mosses. Individual moss species have a physiology which responds best to a combination of conditions, usually provided by the microclimate found in the habitat where that species occurs in its most luxuriant form. Each of the above mentioned factors is briefly discussed in the following sections.

1 LIGHT

The light intensities required to saturate the photosynthetic process in mosses are moderate compared to higher plants and about 70 W/m^2 (10–30 k lux). About 10% of this intensity is all that is required by mosses to reach the light compensation point (i.e. that point at which photosynthetic fixation of CO_2 is just balanced by respiratory production of CO_2). Such values are of rather academic interest. The important thing is to know what intensity is required for what period for a moss plant to be able to recoup during the day respiratory losses during the night. The daily compensation period has been estimated for mosses from tree crowns, in Japan, e.g. *Ulota crispa*, as 2 hours (at optimum light intensities, 80 W/m^2, 20 k lux). Mosses from the tree crown require a brighter light or longer duration to reach a daily compensation period than species such as *Hylocomium caviforme* growing at the base of tree trunks (Hosokawa & Odani 1957; Hosokawa *et al.* 1964). On rainy days or cloudy days, some mosses from tree crowns in Japanese beech forests are scarcely able to compensate for night-time losses. Thus, wherever the particular species is growing, it must receive sufficient light before evaporation leads to a falling water content which will curtail photosynthesis. Clearly excessive radiation will speed water loss and the plants may not be able to compensate for night-time losses before photosynthesis stops. A further factor to be recognized is that the moss must continually replenish the reserves essential for repair of cell membranes during wetting and drying cycles. Thus considerable periods of adequate light and water supply are required by even very slow growing species of poikilohydrous plants and this applies both to mosses and lichens (Smith 1978).

Mosses have been found to be able to adapt their physiology to seasonally changing conditions of water, light and temperature so as to optimize carbon gain (Hicklenton & Oechel 1977). This acclimation seems to be very important but the mechanisms by which mosses and lichens are able to acclimate have yet to be discovered. It has been shown that the radiation levels required to saturate the process of photosynthesis in *Dicranum fuscescens* from northern Canada, at optimum temperature, increase from early in the season until mid summer. A reverse trend occurs late in the season. Early in the season high light intensities occur naturally but the mosses lie under a shallow layer of snow and thus acclimation to photosynthesize optimally at low light levels would be an advantage. Laboratory exposure to high light intensities (120 W/m^2, 30 k lux) early in the season resulted in a reduction of net photosynthesis and samples of *Dicranum fuscescens* from subarctic semi-tundra were more affected than those from a lowland woodland which were presumably acclimated to a higher optimum light intensity. Respectively, reductions of *c.* 30% and 70% of the maximum were found (Hicklenton & Oechel 1976). Studies in Norway on *Dicranum elongatum* and carpet forming mosses have indicated that the continuous incident light experienced by mosses of arctic regions in the midsummer may be deleterious. Abnormal changes were observed in the ultrastructure of the chloroplasts of mosses exposed to continuous light as well as reduced photosynthetic rates (Kallio & Valanne 1975). However, these experiments were done principally with laboratory maintained stocks and acclimation to continuous light may occur naturally in the field. Considerable work has been done to show that lichens are readily able to acclimate to light (Kershaw 1977) and mosses are likely to behave similarly.

2 TEMPERATURE

The sort of questions asked about temperature are: What is the highest and lowest temperatures at which mosses can survive? What are the highest and lowest temperatures at which positive net assimilation occurs given an adequate supply of water and suitable illumination? What is the optimum photosynthetic temperature for a given species? As experimental data has accumulated it has become clear that a simple answer to these questions cannot always be given. For example *Rhacomitrium lanuginosum* from Finland has an optimum for net photosynthesis of +5°C (Fig. 3.1; Kallio & Heinonen 1975) whereas

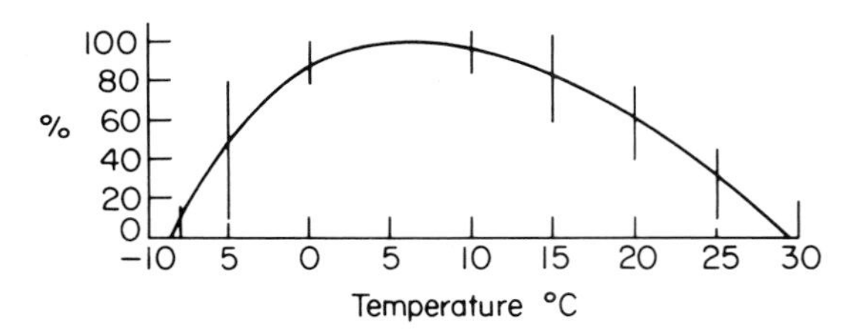

Fig. 3.1 The mean temperature response curve for apparent photosynthesis of nine different strains of *Rhacomitrium lanuginosum*. The photosynthetic activity at +5°C is indicated as 100, the other values representing percentages of it (Kallio & Heinonen 1975).

the same species (Fig. 3.2) from southern England apparently has an optimum close to 30°C (Dilks & Proctor 1975). The situation is further complicated by the observation that material of *Dicranum fuscescens* from the same site in northern Canada has varying temperature optima depending on the season (Fig. 3.3; Hicklenton & Oechel 1976). This will be discussed after a consideration of survival at high and low temperature.

2.1 High temperature

(a) *The high lethal temperature*

For dried moss samples this is between 85 and 110°C (Nörr 1974). As in the case of seeds, it appears that an inverse relationship exists between the logarithm of the observed survival time and the exposure temperature (Roberts & Abdalla 1968; Proctor 1981). In nature, granite mosses and drought tolerant true mosses must withstand high temperatures even in temperate latitudes. Thus, on sand dunes in England, the temperature of a tuft of *Tortula ruraliformis* on a sunny day was recorded as 64·5°C while a colony of *Andreaea rothii* on upland granite rocks reached 50°C (Dilks & Proctor 1976). In Germany, temperatures close to 71°C have been recorded in cushions of *Tortula ruralis*

Fig. 3.2 *Rhacomitrium lanuginosum*, a common moss of acid rocks and peat in montane and arctic regions (photo: M.C.F. Proctor).

and *Rhacomitrium canescens* (Lange 1955). Moist mosses cannot survive such high temperatures and according to most studies do not survive after exposure to between 42 and 47°C although *Rhacomitrium canescens* can withstand 51°C.

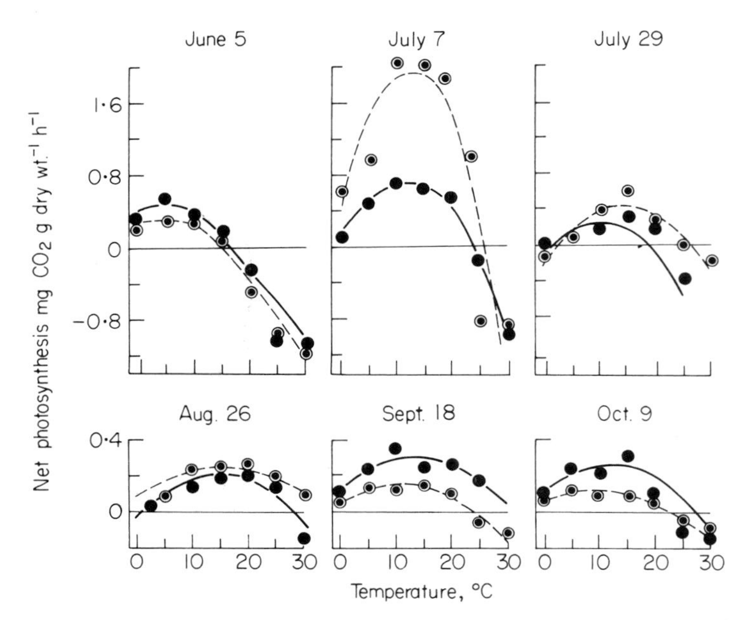

Fig. 3.3 The relationship between net photosynthesis and temperature at saturating light intensity for Timmins (—⊙—) and Mary Jo (—●—) samples of *Dicranum fuscescens*, measured at six periods during 1974 growing season. Values are means of three replicate samples. The curves in each case have been generated from the observed points, using second-degree polynomial regression equations (Hicklenton & Oechel 1976).

(b) *The upper temperature compensation point*

At normal CO_2 concentrations and saturating light intensities, the upper compensation point for mosses such as *Rhacomitrium lanuginosum* and *Dicranum fuscescens* from arctic habitats is about 30°C (Kallio & Heinonen 1975; Hicklenton & Oechel 1976) but many mosses from more temperate climates including *Rhacomitrium lanuginosum* have upper compensation points closer to 40°C (Dilks & Proctor 1975). The aquatic moss *Fontinalis squamosa* has a slightly lower optimum, 30°C, which is expected for a plant growing in the cold water of fast upland streams.

2.2 Low temperature

(a) *The low lethal temperature*

Dry moss samples can frequently survive temperatures below −100°C (i.e. after being placed in liquid nitrogen or liquid air) and in a recent survey eight different genera continued growing or regenerated buds (*Plagiothecium undulatum*) following exposure for six hours to −30°C (Dilks & Proctor 1975). Such survival is only possible when the water content is very low, i.e. less than 30% of the original fresh weight (Malek & Bewley 1978a). Moist mosses are much less temperature tolerant and of the 27 mosses from England, only *Rhacomitrium aquaticum*, *Scorpiurium circinatum* and *Andreaea* spp. showed no evidence of damage following six hours freezing at −10°C. However in boreal and arctic regions, mosses such as *Polytrichum* and *Tortula* grow on embankments which are wetted from snow melt during the day and become encased in ice when the air temperature falls to −20°C or more during the night. In the Canadian prairie provinces, where the temperatures are even colder, *Tortula ruralis* can be encountered in the hydrated state even during the coldest months of the year. An obvious question is how do such mosses survive.

Dr Bewley and co-workers in Alberta, Canada, have made a detailed study of the effects of low temperature on mosses. They work particularly with *Tortula ruralis*. One day, after very fast freezing by plunging into liquid nitrogen, the moss turned brown and loss of chlorophyll was observed. After fast freezing at 60°C/h, there was some degree of browning although the chlorophyll content was the same as controls. Freezing at 3°C/h resulted in no visible symptoms being observed after 24 hours. In the latter two treatments the samples were cooled from +20 to −30°C. It is thus the rate of freezing rather than thawing which causes the greatest damage. *Tortula ruralis* showed no reduction in the ability to synthesize protein from ^{3}H leucine after the following treatment. Samples were frozen slowly to −30°C (the temperature being lowered at 3°C/h) and stored at this temperature for one month, followed by an immersion for ten minutes in liquid nitrogen before being warmed up at 3°C/h. At the cellular level very fast freezing in liquid nitrogen resulted on thawing in the loss of polyribosomes and ability to synthezise protein. Fast frozen samples (at 60°C/h) were also unable to synthesize protein on thawing even though there were comparable quantities of polyribosomes to controls and these *in vitro* could incorporate both ^{14}C leucine and ^{14}C phenylalanine into protein. Following very slow freezing (3°C/h) no abnormal effects were observed on protein synthesis *in vivo* or *in vitro*. On examining the levels of ATP (adenosine triphosphate) in the mosses it was found that after fast freezing and thawing that levels were 60% of controls and declined further during recovery periods. In contrast, after slow freezing and thawing, although ATP levels were only about 40% of controls on thawing, by the time 20°C had been reached, new ATP had been synthesized and normal levels built up after a further 2 hours.

The loss of ATP and leakage of electrolytes (which has also been observed) on slow freezing is evidence of extracellular ice formation leading to desiccation. The survival of samples frozen slowly to −30°C then plunged into liquid nitrogen is evidence that the cells were largely dehydrated. Loss of the ability to synthesise protein on very rapid freezing in liquid nitrogen probably results from loss of compartmentalization within the moss cells due to intracellular ice formation.

Fast freezing injury at 60°C/h is not easily explainable on the basis of intracellular freezing since this should lead to indiscriminate destruction of cell contents. To date, it appears that damage to the ATP synthesizing system occurs. This in turn leads to a cessation of protein synthesis that is essential for the repair of damage to the frozen cells (Malek & Bewley 1978a). Presumably under natural conditions the rate of freezing is such as can be tolerated by most mosses and thus freezing is a hazard to only a small minority of species.

(b) *The low temperature compensation point*

The lowest temperature at which net photosynthesis could be detected in mountain mosses from Romania was −8 to −9°C (in *Brachythecium geheebii*, *Camptothecium phillipeanum* and *Isothecium viviparum*). Respiration continued to rather lower temperatures, −14°C (Atanasiu 1975). Protein synthesis has also been detected at temperatures below freezing (−2·5°C) in *Tortula ruralis* (Malek & Bewley 1978b). Net CO_2 fixation rates a few degrees above 0°C are quite high in most temperate mosses (Fig. 3.1).

2.3 Temperature acclimation

A plant will have a greater chance of survival if it is able to alter its physiology during the growing season so that the temperature for optimum photosynthesis matches that in the environment at any given time. The advantage will be greatest for plants from arctic and subarctic habitats where the growing season is short and where there are large fluctuations in temperature.

Lichens and mosses are able to acclimate to temperature and there is evidence that some higher plants also exhibit this phenomenon

(Hicklenton & Oechel 1976). As mentioned earlier, *Dicranum fuscescens* shows a marked ability to acclimate its temperature for optimum photosynthesis in northern Canada (Fig. 3.3). In addition, variations are evident in the maximum rates of net photosynthesis in different months even though it was found that dark respiration rates showed no seasonal temperature acclimation. The process of acclimation is very rapid taking 2–5 days. Thus samples of this moss placed in day/night temperatures regimes of 18°C/7°C and 8°C/1°C for a period, exhibited respectively photosynthetic optima of 15°C and close to 10°C. Further, the upper temperature compensation point in the plants subjected to the colder regime had also fallen. As with naturally acclimated material, plants acclimated to the lower temperatures showed lower fixation rates than those acclimated to the higher regime. This acclimation is thought to be due to alterations in the activity of enzymes associated with the dark reactions of photosynthesis but little is really known about the mechanism of acclimation in any plants.

In the studies on *Dicranum fuscescens*, the physiological responses were compared between samples from a semi-tundra lowland and samples from a highland site. Both were found to acclimate readily but as shown in Fig. 3.3, they maintained slightly different photosynthetic rates and temperature optima. This suggests that although acclimation provides a considerable degree of physiological plasticity, slight genetic differences between ecotypes may exist and be important.

3 SOLUTE AND PHOTOSYNTHATE MOVEMENT

3.1 Water and solutes

The mechanisms by which ions are initially taken up by mosses will be considered in Chapter 10. Following intracellular absorption, ions are often required at sites distant from where they are taken up, e.g. in the region of new growth or sex organ production. Movement of water can occur from cell to cell via protoplasmic connections; within the cells of special conducting elements; in the cell walls; and along the outer surfaces of the plant. In the morphologically more complex mosses the movement of water and solutes within the plant takes place via the strands of conducting tissue which are evident when cross sections of endohydric moss shoots are observed under the microscope (Zacherl 1956; Bopp & Stehle 1957; Hébant 1977). In mosses like *Polytrichum* (Fig. 3.4) the conducting tissues, the hydrom and leptom, are composed

Fig. 3.4 *Polytrichum commune*, an endohydric moss whose typical habitat is beneath trees on mildly acid soils (photo: M.C.F. Proctor). Note the hairy calyptrae covering the capsules which gives the moss its common name the hairy cap moss.

respectively of two main types of elements, hydroids and leptoids (Fig. 3.5) (Tansley & Chick 1901). As the name suggests, water movement occurs in the hydroids and this is also the pathway for dyes. This

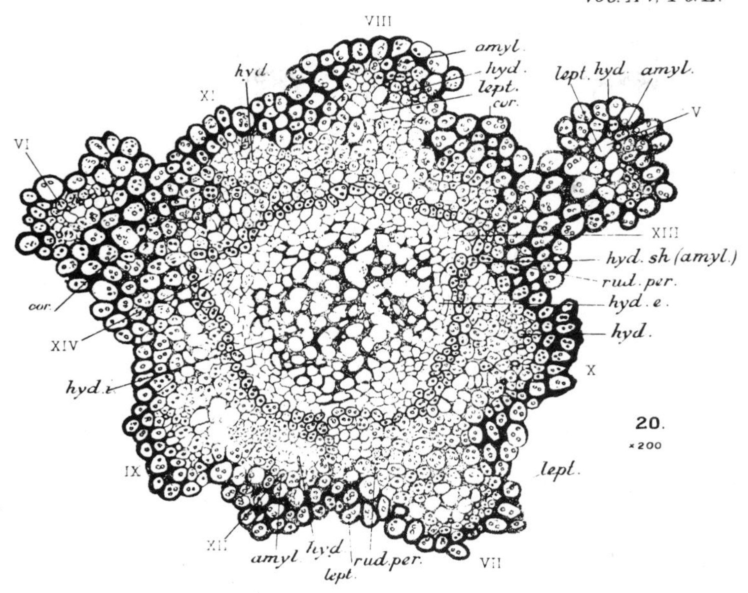

Fig. 3.5 A transverse section of an areal leafy stem of *Polytrichum commune*. (*hyd. i*) hydroids of central cylinder which conduct water; (*hyd. sh.*) hydrom sheath, two layers of starch containing cells; (*lept.*) leptoids which conduct photosynthate; (*cor.*) cortex; (*amyl., hyd, lept.*) tissues of leaf traces i.e. vascular strands running from central cylinder to leaves; (*rud. per.*) rudimentary pericycle (Tansley & Chick 1901).

may be shown by allowing shoots to take up fluorescein or calcofluor white, then sectioning the plants and examining them under a fluorescence microscope (Hébant 1974). The hydroids are elongated cells which normally lack pit connections and lose their protoplasm when mature (Fig. 3.6). The end walls, which become partially hydrolyzed, are inclined and abut on the hydroid immediately above or below (Hébant 1973). The exact path of movement was identified by allowing shoots of *Polytrichum commune* to take up 2% potassium ferrocyanide and then precipitating this as Prussian blue with ferric ions in fixative. Detailed examinations with the electron microscope have shown that the ferrocyanide ion moves through the end walls of the hydroids but not through the lateral walls (Scheirer & Goldklang 1977).

Movement of Pb-EDTA complexes has also been examined in another species *P. juniperinum*. The absorbed lead was precipitated by placing the plants in a hydrogen sulphide environment. The lead was

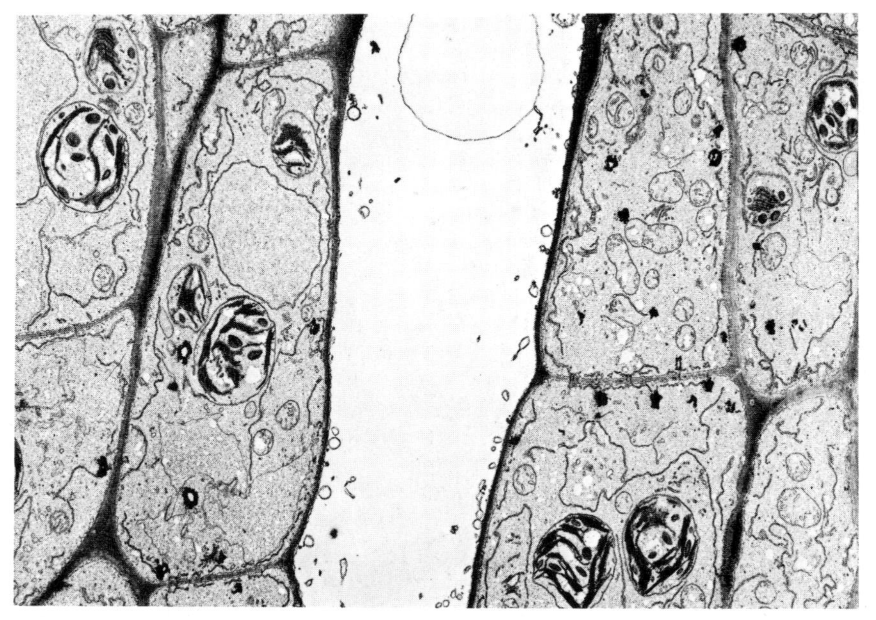

Fig. 3.6 A longitudinal section of a leafy stem of *Polytrichum juniperinum* showing a hydroid of a leaf trace X 5600. This type of cell is thin walled and lacks cell contents (photo: C. Hébant).

found within the lumen of the hydroids confirming this as the movement pathway (Zamski & Trachtenberg 1976).

3.2 Movement of photosynthate and inorganic ions

(a) *Photosynthate*

Sucrose seems to be the main soluble photosynthetic product which accumulates in mosses (Suire 1975; Margaris & Kalaitzakis 1974). This contrasts with the situation in two other poikilohydrous groups of plants, the lichens and liverworts, where sugar alcohols are extremely important (Suleiman *et al.* 1979; Smith 1980). The rate of photosynthate movement in mosses is likely to correlate with the amount of conducting tissue. In ectohydric mosses where this is poorly developed, cell to cell transfer via protoplasmic pit connections is no doubt the principal way by which carbohydrates and other nutrients are transferred from leaves to growing points. In endohydric mosses belonging to the Polytrichaceae there is often a continuous connection of conducting tissue from the leaf to the central tissue of the stem via the leaf

Fig. 3.7 *Hylocomium splendens* an ectohydric moss which occurs in a range of plant communities including woodlands, sheltered grassland, fixed dunes and heaths (photo: M.C.F. Proctor).

trace (Collins & Oechel 1974). Photosynthate moves in the leptoids and not the hydroids of moss conducting tissue and this was shown by applying radioactive bicarbonate solutions to the leaves of *Polytrichum commune* and then following movement of photosynthate by autoradiography (Eschrich & Steiner 1967). Further evidence comes from the observation that if a turgid stem of this moss is carefully cut, a clear exudate may be seen coming from the leptoids which form a zone around the water-conducting central hydroids. This exudate comes from both the leptoids and the associated parenchyma (Hébant 1975).

The leptoids in moss conducting tissue bear strong similarities to sieve elements in phloem of higher plants (Hébant 1975, 1979). They are also extremely similar to the sieve elements of ferns, having slightly oblique end walls with pores and degenerated nuclei at maturity. In addition, when viewed with the electron microscope, many highly refractive spherules are found in both leptoids of mosses and sieve elements in ferns (Behnke 1975). Associated with leptoids are parenchyma cells which, like the companion cells in higher plants or phloem parenchyma in ferns, show increased enzyme activity and are presumably responsible for moving nutrients in and out of the leptoids.

Some mosses, e.g. *Polytrichum*, *Mnium* and *Climacium*, possess effective systems by which to translocate photosynthate to developing buds or new shoots from photosynthesizing leaves or from underground storage 'rhizomes'. The rate of movement of photosynthate from the above ground parts of *Polytrichum alpinum* which grows on the Alaskan tundra, to the underground rhizome system is at least 3 mm/h, i.e. over 7 cm in 24 h (Collins & Oechel 1974). Upward movement appears to occur much faster and rates up to 32 cm/h have been estimated for *Polytrichum commune* (Eschrich & Steiner 1967). In the ectohydric mosses movement is much slower, e.g. in *Hylocomium splendens* (Fig. 3.7) 98% of fixed photosynthate remained at the fixation site after 2 days (Callaghan *et al.* 1978).

(b) *Ions*

Although movement of Pb-EDTA complexes and the ferrocyanide ion takes place in hydroids, a recent study suggests that other ions such as Pb^{2+} and SO_4^{2-} move in the leptoids. Apparently in *Polytrichum juniperinum*, these ions travel across the cortex in the cell walls and accumulate in the leptome (leptoids and associated parenchyma). (Trachtenberg & Zamski 1978). Once in the leptome they are available for translocation to other parts of the plant. In higher plants ions move predominantly in the water conducting tissue (xylem) although some movement of ions no doubt occurs in the photosynthate conducting tissue (phloem). Moss leptoids have affinities, even at the ultrastructural level, with phloem of higher plants (Stevenson 1974). Whether movement of essential cations such as K^+ occurs in the leptoids or in the hydroids of mosses has yet to be determined.

4 MINERAL NUTRITION

The elements needed for the growth of mosses are likely to be those which have been shown essential for the growth of higher plants (Salisbury & Ross 1969). The same combination of nutrients is not always optimal for all stages of the moss life cycle. Thus, one combination of K, Mg and Ca was found to promote the development of rich carpets of protonemata in *Funaria hygrometrica* while others induced much greater gametophore development even though there was restricted protonemal growth (Hoffman 1966).

Endohydric mosses such as *Polytrichum* can take up nutrients from the soil and transport them to the growing apices (see Streeter 1970). Ectohydric species with their poorly developed conducting systems do not transport mineral nutrients a great distance although the fact that the maximum content of N, P, K and Mg occurs in young leaves of *Hyloccmium splendens* during the period of most intense growth suggests that some movement from the older parts of the plant takes place (Skre & Oechel 1979). Ectohydric mosses posses very efficient absorption mechanisms over their whole surface which enable them to take up nutrients directly from water flowing over the surface (see Chapter 10 for more detail). Several studies have suggested that the growth of ectohydric moss carpets may be limited by nutrient supply. For example, the highest rate of production by *Hylocomium splendens* was observed in specimens growing beneath the border of spruce trees where the canopy throughfall was most enriched with tree leachates and dust (see Streeter 1970). In addition laboratory experiments have shown that nutrient deficiencies develop in the ectohydric *Pleurozium schreberi* unless the leaves are sprayed with mineral nutrients. Wefts of the moss growing on soil were unable to take up and transport adequate quantities of nutrients to maintain the growth of the apices. The degree to which carpets of these mosses are nutrient-limited under natural conditions is uncertain. The above study on *Hylocomium* suggests that deficiency is common. However, recent studies in which additional nutrients have been supplied to *Hylocomium splendens, Pleurozium schreberi* and a range of arctic mosses have shown insignificant effects (except in the case of *Sphagnum nemoreum* from a very nutrient poor habitat). It was concluded that mosses are effective at intercepting and absorbing available nutrients and are not normally limited by this factor (Oechel & Sveinbjornsson 1978; Skre & Oechel 1979).

Field observations indicate that certain mosses are associated with acid substrates while others are found in basic situations. In the British Isles *Cratoneuron commutatum* consistently grows in water rich in Ca and its occurrence in water low in this element depends on the presence and ability of Mg to compensate to some extent for the Ca deficiency. However, for many mosses, pH appears to be more important than Ca level. For example, *Acrocladium cuspidatum* can be collected from chalk grassland and fens. Laboratory studies using liquid cultures reveal that the moss grows optimally at a pH of 7·5 and 5 ppm Ca. Higher levels of the element are inhibitory. Media with a pH below 6·0 stop growth even if 5 ppm or more of Ca is present (Streeter 1970).

Recently, interest has been directed towards assessing the importance of bryophytes in the mineral nutrition of natural ecosystems. A study on oak woodlands in Wales (Rieley *et al.* 1979) showed that of K, Ca, Mg and Na, only calcium shows seasonally correlated fluctuations. High levels are observed for this element in summer and winter with lows in spring and autumn. Of the various mosses growing beneath the trees, ecohydric mosses, particularly *Plagiothecium undulatum*, had higher concentrations of Ca, Mg, K and Na (0·33, 0·33, 1·35, 0·11% dry weight for the above moss) than the endohydric species *Polytrichum formosum* (0·14, 0·14, 0·54, 0·05% dry weight respectively). In the Alaskan tundra, higher levels of nitrogen, phosphorus and potassium and greater seasonal fluctuations of nutrients were observed in the ectohydric *Aulacomnium* sp. than in *Polytrichum* (Chapin *et al.* 1980). Ectohydric mosses absorb ions over the whole surface of the gametophore rather than predominantly via the base of the moss cushion. This correlates with recent studies in black spruce forests in Alaska. The green leafy portions of *Sphagnum*, *Hylocomium* and *Pleurozium* had higher phosphate uptake rates than the brown stems to which photosynthate was still translocated. In contrast, the highest phosphate uptake rates in *Polytrichum*, an endohydric species, were observed in the below-ground rhizomes (Chapin pers. comm.).

The source of minerals for the moss carpets in the Welsh oak wood was airborne sea salt and leachates from the treee canopy through which rain falls. *Rhytidiadelphus loreus* removed Ca and K from the throughfall, simultaneously releasing Mg and Na. As might be expected from the observations just mentioned, Ca, Mg, K and Na were not removed in significant amounts from canopy-throughfall that was passed experimentally over leafy shoots of *Polytrichum formosum.* In terms of anions, *Rhytidiadelphus* absorbed NO_3^- from the rain and rather surprisingly *Polytrichum* removed NH_4^+ ions. Leaf litter is another source of minerals for moss carpets and is apparently a more important source of Ca than throughfall. In the above study it was estimated that the combined uptake of Ca, Mg, K and Na by trees and herbs in the wood was 28, 7, 24, 5 kg/ha/yr (respectively) and the total uptake by bryophytes 4, 4, 14, 2 kg/ha/hr. The total input of these ions in the throughfall was 31, 18, 29, 107 kg/ha/yr so that there is an excess of nutrients available to the bryophate community over that needed by them (Rieley *et al.* 1979). In Scandinavian spruce forests, maximum bryophyte growth often occurs beneath the tips of the drooping branches where the rainwater is most enriched.

The degree to which ion-absorption by bryophytes results in the supply for higher plants being insufficient for optimal growth has not been considered in detail. In North Wales it is reported that farmers do not apply phosphate dressings to pastures where there is much moss mixed with the grass because the mosses take most of the phosphate and it is unavailable to other vegetation (Richards 1959).

5 REFERENCES

ATANASIU L. (1975) Photosynthesis and respiration in sporophytes and gametophytes of two species of mosses. *Rev. Roum. Biol.* **20,** 43–8.

BEHNKE H.D. (1975) Phloem tissue and sieve elements in algae, mosses and ferns. In *Phloem Transport*, eds Aronoff S., Dainty J., Gorham P.R., Srivastava L.M. & Swanson C.A., pp. 187–210. Plenum Press, New York.

BOPP M. & STEHLE E. (1957) Zur Frage der Wasserleitung in Gametophyten und Sporophyten der Laubmoose. *Zeitschrift für Botanik* **45,** 161–74.

CALLAGHAN T.V., COLLINS N.J. & CALLAGHAN C.H. (1978) Photosynthesis, growth and reproduction of *Hylocomium splendens* and *Polytrichum commune* in Swedish Lapland. *Oikos*, **31,** 73–88.

CHAPIN T., CHAPIN F.S., JOHNSON D.A. & MCKENDRICK J.D. (1980) Seasonal movements of nutrients in plants of differing growth form in an Alaskan tundra ecosystem: implications for herbivory. *J. Ecol.* **68,** 189–210.

COLLINS N.J. & OECHEL W.C. (1974) The pattern of growth and translocation of photosynthate in a tundra moss, *Polytrichum alpinum. Can. J. Bot.* **52,** 355–63.

DILKS T.J.K. & PROCTOR M.C.F. (1975) Comparative experiments on temperature responses of bryophytes: assimilation, respiration and freezing damage. *J. Bryol.* **8,** 317–36.

DILKS T.J.K. & PROCTOR M.C.F. (1976) Seasonal variation in desiccation tolerance in some British bryophytes. *J. Bryol.* **9,** 239–47.

ESCHRICH W. & STEINER M. (1967) Autoradiographische Untersuchungen zu Stofftransport bie *Polytrichum commune. Planta* (*Berl.*), **74,** 330–49.

GLIME J.M. & Carr, R.E. (1974) Temperature survival of *Fontinalis novae-angliae* Sull. *Bryologist* **77,** 17–22.

HÉBANT C. (1973) Diversity of structure of the water-containing elements in liverworts and mosses, *J. Hattori Bot. Lab.* **37,** 229–34.

HÉBANT C. (1974) Studies on the development of the conducting tissue-system in the gametophytes of some Polytrichales. II. Development and structure at maturity of the hydroids of the central strand. *J. Hattori Bot. Lab.* **38,** 565–607.

HÉBANT C. (1975) The phloem (leptome) of bryophytes. In *Phloem Transport*, eds Aronoff S., Dainty J., Gorham P.R., Srivastava L.M. & Swanson C.A., pp. 211–24. Plenum Press, New York.

HÉBANT C. (1977) *The Conducting Tissues of Bryophytes.* J. Cramer.

HÉBANT C. (1979) Conducting tissues in bryophyte systematics. In *Bryophyte Systematics*, eds Clarke G.S.C. & Duckett J.G. Systematics Association Special Volume, No. 14. pp. 365–83 Academic Press, London.

HICKLENTON P.R. & OECHEL W.C. (1976) Physiological aspects of the ecology of *Dicranum fuscescens* in the subarctic. I. Acclimation and acclimation potential of CO_2 exchange in relation to habitat, light, and temperature. *Can. J. Bot.* **54,** 1104–19.

HICKLENTON P.R. & OECHEL W.C. (1977) Physiological aspects of the ecology of *Dicranum fuscescens* in the subarctic. II. Seasonal patterns of organic nutrient content. *Can. J. Bot.* **55,** 2168–77.

HOFFMAN G.R. (1966) Observations on the mineral nutrition of *Funaria hygrometrica* Hedw. *Bryologist* **69,** 182–92.

HOSOKAWA T. & ODANI N. (1957) The daily compensation period and vertical ranges of epiphytes in a beech forest. *J. Ecol.* **7,** 93–8.

HOSOKAWA T., ODANI N. & TAGAWA H. (1964) Causality of the distribution of corticolous species in forests with special reference to the physio-ecological approach. *Bryologist*, **67,** 396–411.

KALLIO P. & HEINONEN S. (1975) CO_2 exchange and growth of *Rhacomitrium lanuginosum* and *Dicranum elongatum.* In *Fennoscandian Tundra Ecosystems, Part 1. Plants and Microorganisms*, ed. Wielgolaski F.E. *Ecol. Stud.* **16,** 138–48.

KALLIO P. & VALANNE N. (1975) On the effect of continuous light on photosynthesis in mosses. In *Fennoscandian Tundra Ecosystems, Part 1. Analysis and Synthesis*, ed. Wielgolaski F.E. *Ecol. Stud.* **16,** 149–62.

KERSHAW K.A. (1977) Physiological-environmental interactions in lichens. II. The pattern of net photosynthetic acclimation in *Peltigera canina* (L.) Willd. *var. praetextata* (Flöerke in Somm.) Hue, and *Peltigera polydactyla* (Neck.) Hoffm. *New Phytol.* **79,** 377–90.

LANGE O.L. (1955) Untersuchungen uber die hitzerristenz der Moose in Beziehung zu ihrer Verbreitung. 1. Die Resistenz stark ausgetrockneter Moose. *Flora* (*Jena*), **142,** 381–99.

MALEK L. & BEWLEY J.D. (1978a) Effects of various rates of freezing on the metabolism of a drought-tolerant plant, the moss *Tortula ruralis. Plant Physiol.* **61,** 334–8.

MALEK L. & BEWLEY J.D. (1978b) Protein synthesis related to cold temperature stress in the desiccation-tolerant moss *Tortula ruralis. Physiol. Plant.* **43,** 313–19.

MARGARIS N.G. & KALAITZAKIS J. (1974) Soluble sugars from some Greek mosses. *Bryologist*, **77,** 470–2.

NÖRR M. (1974) Hitzeresistenz bei Moosen. *Flora* (*Jena*), **163,** 388–97.

PROCTOR M.C.F. (1981) Physiological ecology of bryophytes. In *Advances in Bryology*, ed Schultze-Motel W., J. Cramer, Germany.

OECHEL W.C. & SVEINBJORNSSON B. (1978) Primary production processes in arctic bryophytes at Barrow, Alaska. *Ecol. Stud.* **29,** 269–98.

RICHARDS P.W. (1959) Bryophyta. In *Vistas in Botany*, ed. Turrill W.B., pp. 387–420. Pergamon Press, London.

RIELEY J.O., RICHARDS P.W. & BEBBINGTON A.D.L. (1979) The ecological role of bryophytes in a North Wales woodland. *J. Ecol.* **67,** 497–528.

ROBERTS E.H. & ABDALLA F.H. (1968) The influence of temperature, moisture, and oxygen on period of seed viability in barley, broad beans, and peas. *Ann. Bot.* (*Lond.*), **32,** 97–117.

SALISBURY F.B. & ROSS C. (1969) *Plant Physiology*. Wadsworth Publishing Co., Belmont, California.

SCHEIRER D.C. & GOLDKLANG I.J. (1977) Pathway of water movement in hydroids of *Polytrichum commune* Hedw. (Bryopsida). *Am. J. Bot.* **64,** 1046–7.

SKRE O. & OECHEL W.C. (1979) Moss production in a black spruce *Picea mariana* forest with permafrost near Fairbanks, Alaska, as compared with two permafrost-free stands. *Holarctic Ecol.* **2,** 249–54.

SMITH D.C. (1978) What can lichens tell us about real fungi. *Mycologia,* **70,** 915–34.

SMITH D.C. (1980) The Lichen Symbiosis. In *Cellular Interactions in Symbiotic and Parasitic Associations*, eds Cook C.B., Pappas P.W. & Rudolph E.D. 5th Ann. Colloq. Coll. Biol. Sci., Ohio State Univ. Press.

STEVENSON D.W. (1974) Ultrastructure of the nacreous leptoids (sieve elements) in the polytrichaceous moss *Atrichum undulatum*. *Am. J. Bot.* **61,** 414–21.

STREETER D.T. (1970) Bryophyte ecology. *Sci. Prog. Oxf.* **58,** 419–34.

SUIRE C. (1975) Les donnees actuelles sur la chimie des bryophytes. *Rev. Bryol. Lichenol.* **41,** 105–256.

SULEIMAN A.A.A., BACON J., CHRISTIE A. & LEWIS D.H. (1979) The carbohydrates of the leafy liverwort *Plagiochila asplenioides* (L.) Dum. *New Phytol.* **82,** 439–48.

TANSLEY A.G. & CHICK E. (1901) Notes on the conducting tissue system in Bryophyta. *Ann. Bot.* (*Lond.*), **15,** 1–38.

TRACHTENBERG S. & ZAMSKI E. (1978) Conduction of ionic solutes and assimilates in the leptom of *Polytrichum juniperinum* Willd. *J. Exp. Bot.* **29,** 719–27.

ZACHERL H. (1956) Physiologische und okologische Untersuchungen uber die innere Wasserleitung bei Laubmoosen. *Zeitschrift für Botanik*, **44,** 409–36.

ZAMSKI E. & TRACHTENBERG S. (1976) Water movement through hydroids of a moss gametophyte. *Israel J. Bot.* **25,** 168–73.

CHAPTER 4
SEX AND CYTOGENETICS

The life cycle of mosses was outlined in the first chapter of this book and illustrated in Fig. 1.7. The regular alternation of the green leafy haploid gametophyte generation with the capsule bearing diploid sporophyte generation has long intrigued botanists and naturalists partly because both generations are so conspicuous. Several theories have been advanced to explain why the fertilized egg cell within the archegonium should develop into a capsule while the spore within that capsule develops into a plant with leafy shoots (Bauer 1967).

At first sight, one might suppose that the number of chromosomes controls the outcome of the development of egg or spore. However this is not so since sporophytes can be produced experimentally direct from protonemata in *Pottia intermedia* and *Splachnum ovatum* (Bopp 1968) or from unfertilized egg cells (apogamy) in a number of mosses. Further, some mosses are polyploid having several sets of chromosomes even in the gametophyte stage (see section 4c). Conversely, gametophytes can be induced to form from sporophyte tissue (apospory) (see Chapter 6, section 2.2e). Traditionally it has been stated that the environment in which the fertilized egg (the first cell of the sporophyte generation) develops is very different from that in which the spore (the first cell of the gametophyte generation) germinates and in consequence quite separate sets of genes are activated. These would then determine the pattern of the resulting growth. However studies on ferns do not support this idea because very young developing sporophytes excised from archegonia and placed on nutrient media still form sporophytes (Bell 1970). A further theory holds that the act of fertilization results in the introduction of factors from the sperm cytoplasm which trigger the genes that will control morphogenesis. This too seems unlikely since the sperm cytoplasm is rapidly digested by the egg.

There is now evidence that the ability to form a sporophyte is programmed into a moss cell (usually the egg) during its differentiation. If fern gametophytes with developing archegonia are fed thiouracil the eggs, when later fertilized, grow abnormally producing gametophyte-like tissue. These often bear structures at the margin resembling antheridia (Bell 1970). Thiouracil is a compound which interferes with the synthesis of ribonucleic acid (RNA) and consequently with protein and enzyme synthesis. It is supposed that treatment of the developing archegonia inhibits the activation, in the eggs, of genes which determine that the destiny shall be a sporophyte pattern of growth. In the case of sporophytes developed on protonemata, low sugar and high nitrogen enhance development, as does chloral hydrate, but the mode of action of these initiating factors on apical cells of the protonema is unknown (Bopp 1968).

Cytochemical and ultrastructural studies emphasize that the difference between the egg and spore lies within these structures. The former is several times as large and is rich in organelles, particularly ribosomes, as well as in nucleic acids and proteins.

The current hypothesis requires, as its corollary, that in the production of spores, the pattern of gene activation reverts to a state which will lead to gametophyte development. Some evidence that this occurs is derived from studies on lily pollen during the early stages of meiosis. This process is also involved in the formation of moss spores within the capsule. In lily pollen, there is a marked reduction during early meiosis of the amounts of RNA and in the number of ribosomes within the cytoplasm. This is exactly the opposite to what is observed during

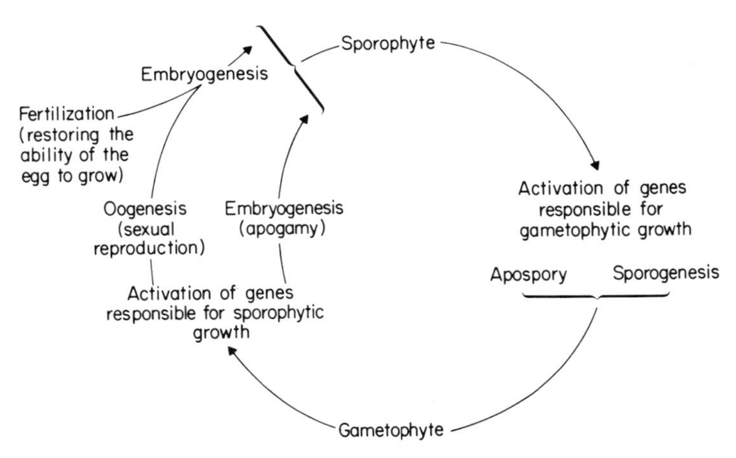

Fig. 4.1 The moss life cycle interpreted as resulting from the alternate activation of two different sets of genes. The first set act during spore formation (sporogenesis) or on experimentally induced sporophyte tissue to produce protonemata (apospory). The other set come into play during development of the egg (oogenesis) or in the formation of apical cells on experimentally induced protonema which in consequence form sporophytes directly (apogamy) (Bell 1970).

egg development within archegonia. In summary, Fig. 4.1 illustrates a likely way in which alternation of generations, a key feature of the moss life history, is determined at various stages in the cycle (Bell 1970).

The following sections in this chapter discuss the various stages in the moss life cycle from the initiation of sex organ development to fertilization of the egg within the archegonium.

1 DEVELOPMENT OF SEX ORGANS

The conditions which induce the development of antheridia and archegonia on the leafy gametophores of mosses have been studied in very few species (Hallet 1974). Only *Sphagnum plumulosum* is known to react to day length. It remains sterile if grown under long-day conditions (16 hours illumination in 24 hours) but develops sex organs when the days are short (8 hours) (Benson-Evans 1964). This correlates with field observations that fertile plants occur from February to March in the U.K. In the U.S.A. the antheridia of *S. subsecundum* develop in summer although release of the sperm cells does not usually occur until the following spring (Bryan 1915). In other mosses so far studied, i.e., *Polytrichum aloides*, *Funaria hygrometrica* and *Leptobryum pyriforme* the length of day is unimportant for sex-organ initiation (Benson-Evans 1964; Monroe 1965; Chopra & Rawat 1977).

In most mosses it would seem that temperature is the key factor controlling the development of antheridia and archegonia. Sex organs are formed in *Polytrichum aloides* at 21°C but not at 10°C even though gametophore growth is better at the latter temperature. The detailed study on *Leptobryum pyriforme* showed that this species remains sterile if grown at 25°C. It is a convenient moss for experimental work as it will form gametophores with sex organs in as little as five weeks from the time of spore germination. Gametophores transferred from 25°C to 10°C became abundantly fertile after 8 days. At 15°C a large number of shoots also developed sex organs but this took longer (11 days). Only few gametophores became fertile on transfer to 20°C even after 30 days (Chopra & Rawat 1977). *Funaria hygrometrica* also seems to need a low temperature, around 10°C, for sex organ initiation (Monroe 1965). In *Leptobryum pyriforme* both antheridia and archegonia are borne on the same shoot but the antheridia develop first. Once the antheridia have been formed by exposure of the plants to low temperature, archegonia will form even though inductive temperatures are not maintained.

Sex organs cannot be induced to develop by exposing *Leptobryum pyriforme* to low temperature in the dark but low light intensities will suffice (9 W/m², 2 k lux). At high temperatures, 20°C, 70 W/m² (16 k lux), are needed to induce fertility. Growing this moss in the presence of a range of nutrients and plant hormones failed to induce sex organ development so that the hormone controlling sex organ development is still unknown (Chopra & Rawat 1977).

2 FERTILIZATION

2.1 RELEASE OF SPERMS

The antheridia of true mosses are large club shaped organs consisting of a stalk, jacket and tip which enclose a sperm chamber (Fig. 4.2). Clusters of antheridia develop on the leafy gametophore interspersed with sterile protective hairs, the paraphyses. Sometimes special leaves, the perigonial leaves, surround these structures and play an important part in sperm dispersal (See p. 54).

In *Polytrichum*, *Atrichum* and *Mnium*, fluid is secreted into the base of the antheridium below the sperm mass as the antheridium matures. This is readily seen in the first two genera but difficult to observe in *Mnium* before the main sperm mass has been extruded. The fluid in the base acts as a hydraulic ram being responsible for the build up of pressure within the antheridium which stretches the wall and thus stores potential energy. This is converted into kinetic energy as the antheridium opens.

The sequence of events is as follows: the outer walls of the cap cells on the top of the antheridium absorb water, swell and cause the inner walls to burst inwards depositing cap-cell cytoplasm on the top of the sperm mass (Hausmann & Paolillo 1978). The antheridium is under considerable pressure and the softening outer walls of the cap cells cannot resist this for long. The walls rupture and the sperm mass is forced out of the orifice in less than 20 seconds by the rapid contraction of the stretched jacket. Any part of the sperm mass which fails to exit in this way is extruded more slowly by uptake of water into the fluid at the base of the sperm mass. It usually takes less than 4 minutes for the antheridium to empty completely. In *Polytrichum* 100% of the sperms are frequently ejected during the rapid contraction phase whereas in the other genera at least part remains for release by the slow phase (Paolillo 1977a; Hausmann & Paolillo 1977).

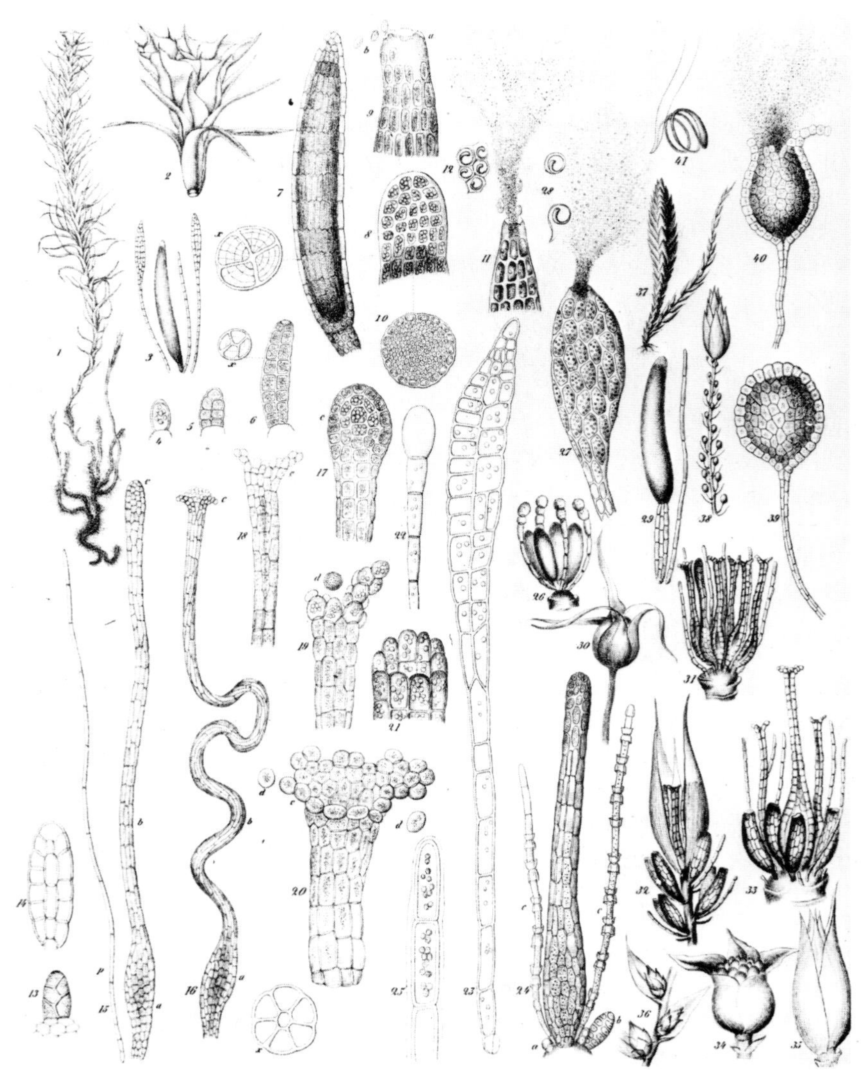

Fig. 4.2 Antheridia and archegonia as seen with the light microscope (Schimper 1860) (1) male plant of *Polytrichum juniperinum*, (2) details of antheridial head, (3) antheridia and paraphyses, (4, 5, 6) developing antheridia, (7, 8, 9, 10) details of the antheridium, (11) the antheridium dehiscing, (12) a sperm cell, (13, 14, 15, 16) developing archegonia, (17, 18, 19, 20, 21) details of the tip of the archegonia and its opening, (23) a paraphysis, (24) archegonia of *Diphyscium foliosum*, (25) a paraphysis tip, (26) antheridia and paraphyses of *Funaria hygrometrica*, (27) dehiscence of an antheridium, (28) a sperm cell, (29) antheridia of *Timmia megapolitana*, (30) antheridial head of *Splachnum gracilis*, (31) archegonia of *Bryum pseudotriquetrum* (32) archegonia and antheridia of *Pohlia nutans* (33) antheridia and archegonia of *Bryum bimum*, (34)

In *Funaria hygrometrica* the tip cells of the antheridia break down as in the other mosses and this is again followed by a two phase release mechanism. However fluid is absent from the base of the antheridia and the effectiveness of the rapid release phase is entirely the function of contraction of the jacket wall distended during maturation of the sperm mass. Not all the sperms are extruded during this contraction. The slow release phase is also different in *Funaria hygrometrica.* Instead of the remaining sperm mass being forced out, it is pulled out of the antheridium by the cohesiveness of the larger extruded portion of the sperm mass which progressively swells to more than five times its volume in the unopened antheridium (Paolillo 1977b).

The antheridia of Sphagnopsida release sperms in a totally different manner from the other mosses. Cells at the apex absorb water and the antheridium opens by a number of irregular lobes which bend backwards. The membranes of the sperm mother cells which make up the sperm mass rupture and the sperms swim away through the open apex of the antheridium (Muggoch & Walton 1942; Parihar 1977).

2.2 Dispersal of sperms

If any part of an extruded sperm mass of e.g. *Polytrichum,* touches an air–water interface, the whole sperm mass disperses immediately. A monolayer of sperm-mother cells is formed on the surface of the water and the sperm cells liberate themselves from the gelatinous shell of mucilage formed from the remains of cells (Lal & Bell 1975). This also applies to the cohesive sperm-mass of *Funaria.* Evidently, the sperm masses of these mosses contain substances that lower surface tension. These have been identified as being of lipid nature. In *Polytrichum* and *Atrichum* the fats are mainly sterols although triglycerides also occur. Analysis of the sperm mass of *Mnium* showed the presence of a diglyceride and a substance which may be a monoglyceride. The swelling of the sperm mass in *Funaria,* and that which also occurs in *Mnium,* after extrusion from the antheridium may be due to the presence of these monoglycerides. They have the capacity of swelling in aqueous environments (Paolillo 1979). It is interesting that tests with a number of

---

antheridial head of *Mnium* sp., (35) gemmae of *Bryum,* (36) gemmae of *Hypnum,* (37) antheridia bearing branch of *Sphagnum acutifolium,* (38) antheridia exposed by dissecting off leaves, (39) a mature antheridium, (40) dehiscence of the antheridium (41) a sperm cell.

staining reagents failed to identify fats in the sperm mass of *Sphagnum* sp. This correlates with the mechanism of sperm release previously described. Thus, in *Sphagnum*, dispersal of the sperms to the vicinity of the archegonia must be effected by the activity of the sperm itself or by passive transport via water seepage through the saturated peat moss clump (Muggoch & Walton 1942).

2.3 Transfer of sperms to the Archegonium

Sperm cells of mosses are spirally coiled with two long flagella at one end (Fig. 4.2) and their ultrastructure has been the subject of recent studies. These have shown details of the flagella and that sperm cells contain a single plastid filled with starch as well as a nucleus whose size is related to the amount of contained DNA (Duckett & Carothers 1979; Robenek 1979). Moss sperm cells, like those of liverworts, have been shown to maintain their mobility for up to six hours. The distance swum in this time is very limited. For example sperms released at one end of a pool of water 1 cm long and 0·5 cm wide were, with the exception of one or two individuals, still crowded after an hour near the end where they were liberated (Muggoch & Walton 1942). The transfer of sperms to the vicinity of the archegonia must therefore occur in other ways. If antheridia burst onto a water-covered moss tuft composed of gametophores with both kinds of sex organs, the reduction in surface tension due to lipids in the sperm mass may be sufficient to transfer the sperms to the vicinity of the archegonium though it is not known how they cross the interface from the oil to the water. Alternatively water dripping from branch to branch or seeping through moss cushions on mires may carry the sperms passively. Some mosses such as *Polytrichum* have evolved a splash-cup mechanism which is an adaptation to increase the distance of sperm dispersal. The antheridia are surrounded by a whorl of stiff perigonial leaves which form a cup that may be as much as 6mm in diameter (see Fig. 1.4). Under wet conditions antheridia discharge masses of sperms into the cup. Small raindrops splash onto and then out of the cup carrying sperm cells many centimetres away (Brodie 1951). The leaves of the gametophores bearing archegonia point vertically so that the sperm carrying drop is caught, and the water then runs down between the leaves to where the archegonia are situated. In general, the distance of effective sperm dispersal via splash cups depends on their size and elevation. In large mosses, e.g. *Dawsonia superba* the distance of sperm dispersal has been estimated as from 1·5 to 2·0 m both from field observations and laboratory studies. In this moss the splash cup mechanism is thought to account for almost all fertilizations (Clayton-Greene *et al.* 1977). In smaller mosses such as *Mnium ciliare*, the maximum distance observed in the field between a sporophyte (which had developed from a fertilized archegonium) and the nearest male plant was about 5 cm. However laboratory studies showed that sperms could be splashed more than twice as far on a significant number of occasions and sometimes close to 50 cm (Reynolds 1980). It is also possible that, in some mosses, insects may carry sperms to the archegonia. Oribatid mites, springtails, small midges, leaf hoppers, aphids and spiders have been observed to visit the antheridial and archegonial heads of *Polytrichum*. The insects can become smeared with sperm cells but are apparently attracted by the abundant mucilage exuded by the paraphyses which helps to keep the sex organs moist (Harvey-Gibson & Miller-Brown 1927). In *Bryum capillare*, female plants 2 m from male plants were not fertilized if covered with a fine net which excluded insects but were fertilized when the net was removed and insects could visit the plants (Gayat 1897). The importance of insect mediated sperm transfer has not been assessed recently. The antheridial heads of the Polytrichales are often conspicuously coloured red or yellow, but they have never been shown to attract insects in the same regular way as the flowers of higher plants. It is not known why the perigonial leaves surrounding the antheridia in *Polytrichum* are brightly pigmented whereas those surrounding the archegonia are the normal green colour. If attraction to animals for sperm dispersal was the primary function one would have expected evolution or pigmentation in leaves surrounding both types of sex organ.

In spite of the various possibilities for sperm transfer, it seems to be a rather inefficient process. This may account for the high proportion of monoecious mosses (i.e. both archegonia and antheridia are borne on the same plant). Almost 50% of the mosses from the British Isles exhibit this condition (Longton 1976). A strange situation occurs in the genus *Dicranum*, which is dioecious (male and female sex organs on different plants). In those species which fruit most commonly, male plants of minute size are found attached to leaves or tomentum of the larger female plant. Apparently a hormone is released by the female plant and this induces any *Dicranum* spores landing on it to develop into a dwarf male. By this means both outbreeding and sperm transfer are facilitated (Crum 1976). Dwarf males have been discovered in one or more species of some sixty genera of mosses (Ramsay 1979). However, in the tropical mosses *Macromitrium* and *Schlotheimia*, a further development has occurred. In some species in these genera, spores of two sizes are produced in, and distributed from, a single capsule. The larger spores germinate and give rise to large gametophores bearing archegonia. The small spores form dwarf males developing on the leaves of the female plants. In this case, although the size of the plant is genetically determined rather than due to the influence of a hormone produced by the female, it is still possible that the latter may exude substances which stimulate germination of small spores of the same species and inhibit those of other mosses (Ramsay 1979).

In normal dioecious mosses, the distance for successful sperm transfer between male and female plants is quite limited. Thus no capsules were found in *Pleurozium schreberi* and *Climacium dendroides* if the archegonia bearing plants were more than 10 cm from the male plants. Mosses with well developed splash cups fare a little better. Capsules were recorded up to 75 cm from the antheridial heads of *Polytrichum juniperinum* from the subarctic (Longton 1976). *Sphagnum subsecundum* may have evolved an answer to this problem by occasionally foregoing the advantage of gene transfer through fertilization. In this moss, fusion of the egg cell in the archegonium with the adjacent neck canal cell sometimes occurs obviating the need for a sperm (Bryan 1920).

2.4 Fertilization

When the archegonia are mature (Fig. 4.2) the cap cell bursts open and the neck canal cell breaks down so opening a pathway for fertilization of the egg cell. Sperm cells arriving in the vicinity of the archegonial neck are strongly attracted and swim towards the canal. The attractant substance in mosses seems to be sucrose; experimentally it was found that sperm cells of *Funaria hygrometrica*, *Leptobryum pyriforme* and *Brachythecium rivulare* were attracted by 0·001–0·1% sucrose but nothing else (Machlis & Rawitscher-Kunkel 1967; Garjeanne 1932). The sperm nucleus has a high degree of flexibility and it is thought that this enables sperm to traverse the long narrow neck of the archegonium. The neck is filled with mucilage and flagella would be quite ineffective in providing the necessary driving force in this narrow column of viscous fluid (Lal & Bell 1975). Eventually the sperm cell reaches and fertilizes the egg cell at the base of the archegonium. Moss eggs are unique in that at the time of fertilization they appear to float free in the base of the archegonium (Sarafis pers. comm.). The fertilized egg cell grows and divides producing a hormone which inhibits development of other surrounding archegonia. As a result, a single capsule grows from the apex of each gametophore but occasionally several capsules are found. Presumably this is due either to more than one archegonium being fertilized simultaneously so that they are at the same developmental stage or, to a lower production of the inhibiting hormone by the first archegonium fertilized. The latter would seem to pertain for such species as *Plagiomnium insigne*, a moss from the forest floors of western North America, as this moss often has nine or more capsules growing from the apex of a single leafy shoot. In Europe, *Plagiomnium undulatum* may have 5–6 capsules per leafy shoot. (Lacey 1950) and *Dicranum majus* may also be polysetous (Sowter 1948).

3 CYTOGENETICS

Sperms of one moss species are often attracted by archegonia of a second and may even penetrate the egg. Fertilization will not occur unless the two species are closely related. Natural hybrids are found occasionally between related species or even between related genera, providing that the chromosome complements are similar. Hybrids are also more likely when gametophores of the parent species are morphologically similar. Such crosses are identified by the abnormal capsules which develop following fertilization.

Both naturally occurring and laboratory induced hybrids will form between the three genera *Astomum*, *Hymenostomum* and *Weissia*. The capsules of the first genus break open irregularly at maturity, those of the second have a functional lid (operculum) while those of the third normally have both lid and peristome. When archegonia of e.g. *Astomum ludovicianum* are fertilized with sperm cells of *Weissia controversa*, capsules form which are intermediate between the parents. However, the capsules contain largely shrunken spores and all are infertile. It is found that in such hybrids, chromosomes pair and segregate with no difficulty during spore formation within the capsules, but the spores fail to grow properly. Incompatibility seems therefore to be at the level of the gene rather than inability of the chromosomes to associate. Thus even though these three genera are clearly closely related, there are effective barriers which prevent loss of generic identity by hybridization (Anderson & Lemmon 1972). Hybrids can also be produced between *Physcomitrella*, *Physcomitrium* and *Funaria* which show similar gradations in capsule complexity.

In *Astomum*, *Hymenostomum* and *Weissia* the basic number of chromosomes is 13 although, as in several other mosses, there may be a number of diminutive chromosomes termed m-chromosomes. Populations of *Weissia controversa* containing m-chromosomes are common in warm dry habitats in the U.S.A. m-Chromosomes are less common or absent in populations from wetter or colder habitats. It is hypothesized that m-chromosomes provide one way for an increased capacity for variability in mosses through the development of new experimental genotypes. Normally these only survive in populations growing in very favourable conditions but are especially valuable for mosses which propagate vegetatively and have a low outbreeding capability (Anderson & Lemmon 1972). The m-chromosomes pair loosely during meiosis, prior to the formation of the spores in the capsule, but behave unlike the normal chromosomes. However at least one m-chromosome is regularly distributed to each of the four daughter nuclei at the end of the meiotic process. Diminutive supernumerary chromosomes often called B-chromosomes are also found in some flowering plants and the grasshopper *Myrmeleotettix maculatus* (Anderson & Lemmon 1972).

Another unusual feature of mosses is that one of the chromosomes of the regular complement is often larger than all the others and is sometimes called the H-chromosome as it is heterochromatic. Heterochromatin is however not limited to such chromosomes and recent studies suggest that the term should be abandoned (Newton 1979).

The consequence of a free-living haploid gametophyte is that all the genes on the chromosomes are functionally dominant. Thus at first sight, moss gametophores cannot possess recessive alleles that are selected against under existing conditions, but which could later be advantageous under changing conditions. However if one or more of the chromosomes are duplicated (even to the extent of several complete sets being present) then the above disadvantage can be circumvented. Duplication of chromosomes is common in mosses. Thus while the basic chromosome number is $n = 7$ (Smith 1978; Newton 1979) about 57% of Bryopsida have $n = 10$ to 14 (Longton 1976). About 20% of true mosses are clearly polyploid, i.e. having more than one complete set of chromosomes in the gametophyte generation. Granite mosses have 11 chromosomes while in peat mosses the number is 21 (19 + 2 m). However, races occur of many *Sphagnum* species which contain 42 (38 + 4) (Wylie 1957). Some of the true mosses such as *Tortula muralis* are highly polyploid with populations having $n = 28$ to 66 (Puri 1973).

In addition to shielding potentially useful alleles, polyploid plants tend to be larger showing an increase in size of organs and cells. However a correct balance is also important. Thus polyploids of *Physcomitrium pyriforme* have more chloroplasts per cell but a slower growth rate. It has been suggested that the increase in cell size is proportionally less so that the chloroplasts are crowded, presumably reducing their efficiency (Lewis 1961). Occasionally diploids (having two sets of chromosomes) and tetraploids (with four sets of chromosomes) look different enough that they are described as separate species, e.g. *Fissidens cristatus* and *F. adianthoides*, or given varietal status as in *Atrichum undulatum* (*var. minus*, $n = 7$; *var. undulatum*, $n = 14$; *var. haussknechtii*, $n = 21$). In most mosses, however, diploid and tetraploid races of the same species are morphologically indistinguishable as in *Tortula muralis* or *Funaria hygrometrica* where the chromosome number may be 14, 28 or 56 (Anderson 1974; Puri 1973).

4 REFERENCES

ANDERSON L.E. & LEMMON B.E. (1972) Cytological studies of natural intergeneric hybrids and their parent species in the moss genera, *Astomum* and *Weissia*. *Ann. Missouri Bot. Gard.* **59**, 382–416.

ANDERSON L.E. (1974) Bryology 1947–1972. *Ann. Missouri Bot. Gard.* **61**, 56–85.

BAUER L. (1967) Determination von Gametophyt und Sporophyt. *Handbuch der Pflanzenphysiologie*, **18**, 235–56.

BELL P.R. (1970) The archegoniate revolution. *Sci. Progr. Oxf.* **58**, 27–45.

BENSON-EVANS K. (1964) Physiology of the reproduction of Bryophytes. *Bryologist*, **67**, 431–45.

BOPP M. (1968) Control of differentiation in fern allies and bryophytes. *Ann. Rev. Pl. Physiol.* **19**, 361–77.

BRODIE H.J. (1951) The splash dispersal mechanism in plants. *Can. J. Bot.* **29**, 224–34.

BRYAN G.S. (1915) The archegonium of *Sphagnum subsecundum*. *Bot. Gaz.* **49**, 40–56.

BRYAN G.S. (1920) The fusion of the ventral canal cell and egg in Sphagnum *subsecundum*. *Am. J. Bot.* **7**, 223–30.

CHOPRA R.M. & RAWAT M.S. (1977) Studies on the initiation of sexual phase in the moss *Leptobryum pyriforme*. *Beitr. Biol. Pflanz.* **53**, 353–7.

CLAYTON-GREENE K.A., GREEN T.G.A. & STAPLES B. (1977) Studies of *Dawsonia superba*. 1. Antherozoid dispersal. *Bryologist*, **80**, 439–44.

CRUM (1976) Mosses of Great Lakes forest Region. Revised ed. *Univ. Mich. Herb.* 1–404.

DUCKETT J.G. & CAROTHERS, Z.B. (1979) Spermatogenesis in the systematics and phylogeny of the Musci. In *Bryophyte Systematics*, eds Clarke G.C.S. & Duckett J.G. Systematics Association, special volume No. 14, pp. 385–423, Academic Press, London.

GARJEANNE A.J.M. (1932) Physiology. In *Manual of Bryology*, ed. Verdoorn F., pp. 207–32. Martinus Nijhoff, The Hague.

GAYAT L.A. (1897) Recherches sur le developpement de l'archegore chez les muscinees. *Ann. Sci. Nat. Ser.* **8, 3,** 161–258.

HALLET J. (1974) Morphogenese du gamétophyte feuillé du *Polytrichum formosum* Hed. 11. étude histochimique, histoautoradiographique, et cytophotométrique de la région apicule pendent la phase reproductrice. *Ann. Sci. Nat. Bot. Paris Ser 12,* **15,** 321–88.

HARVEY-GIBSON R.J. & MILLER-BROWN D. (1927) Fertilization of Bryophyta. *Ann. Bot. (Lond.),* **49,** 190–1.

HAUSMANN M.K. & PAOLILLO D.J. (1977) On the development and maturation of antheridia in *Polytrichum. Bryologist,* **80,** 143–8.

HAUSMANN M.K. & PAOLILLO D.J. (1978) The tip cells of antheridia of *Polytrichum juniperinum. Can. J. Bot.* **56,** *1394-9.*

LACEY W.S. (1950) Notes on fruiting of *Mnium undulatum. Trans. Br. Bryol. Soc.* **1,** 370–3.

LAL M. & BELL P.R. (1975) Spermatogenesis in mosses. In *The Biology of the Male Gamete,* eds Duckett J.G. & Racey P.A. Supplement No. 1 Biol. J. Linn. Soc. vol. 7, pp. 85–95. Academic Press, London.

LEWIS K.R. (1961) The genetics of bryophytes. *Trans. Br. Bryol. Soc.* **4,** 111–30.

LONGTON R.E. (1976) Reproductive biology and evolutionary potential in Bryophytes. *J. Hattori Bot. Lab.* **41,** 205–23.

MACHLIS L. & RAWITSCHER-KUNKEL E. (1967) Mechanisms of gametic approach in plants. In *Fertilization,* eds Metz C.B. & Monroy A., pp. 117–61. Academic Press, New York.

MONROE J.H. (1965) Some factors evoking formation of sex organs in *Funaria. Bryologist,* **68,** 337–9.

MUGGOCH H. & WALTON J. (1942) On the dehiscence of the antheridium and the part played by surface tension in the dispersal of spermatocytes in Bryophyta. *Proc. R. Soc. Lond. B. Biol. Sci.* **130,** 448–61.

NEWTON M.E. (1979) Chromosome morphology and bryophyte systematics. In *Bryophyte Systematics,* eds Clarke G.S.C. & Duckett J.G. Systematics Association, special volume No. 14, pp. 207–29. Academic Press, London.

PAOLILLO D.J. (1977a) On the release of sperms in *Atrichum. Am. J. Bot.* **64,** 81–5.

PAOLILLO D.J. (1977b) Release of sperms in *Funaria hygrometrica. Bryologist,* **80,** 619–24.

PAOLILLO D.J. (1979) On the lipids of the sperm masses of three mosses. *Bryologist,* **82,** 93–6.

PARIHAR N.S. (1977) Bryophata. In *An Introduction to Embryophyta, Vol. I,* 5th ed. Central Book Depot, Allahabad.

PURI P. (1973) *Bryophytes, a Broad Perspective.* Atma Ram, Dehli.

RAMSAY H.P. (1979) Anisospory and sexual dimorphism in the Musci. In *Bryophyte Systematics,* eds Clarke C.G.S. & Duckett J.G. Systematics Association, special volume No. 14, pp. 281–316. Academic Press, London.

REYNOLDS D.N. (1980) Gamete dispersal in *Mnium ciliare. Bryologist,* **83,** 73–7.

ROBENEK H. (1979) Die Feinstruktur der Spermatosoiden von Polytrichum piliferum unter besonderer Berucksichtigung der Plasmalemmadifferenzierung wahrend der spermatogenese. *Ber. Dtsch. Bot. Ges.* **92,** 595–608.

SCHIMPER W.P. (1860) *Icones morphologicae atque organographicae introductionem synopsis muscorum europaeorum.* E. Schweizerbart, Stuttgartiae.

SMITH A.J.E. (1978) Cytogenetics, biosystematics and evolution in the Bryophyta. In *Advances in Botanical research,* ed. Woolhouse H. Vol. 6, pp. 196–277. Academic Press, London.

SOWTER F.A. (1948) The polysetous inflorescence of *Dicranum majus* L. *Trans. Br. Bryol. Soc.* **1,** 73–4.

WYLIE A.P. (1957) The chromosome numbers of mosses. *Trans. Br. Bryol. Soc.* **3,** 260–78.

CHAPTER 5
CAPSULES AND SPORE DISPERSAL

1 DEVELOPMENT OF THE SPOROPHYTE

The time required for the development of a mature capsule from the egg within the archegonium on the gametophore varies with the species of moss (Bauer 1963). In temperate zones, at least three patterns have been recognized: 1 a 6 month development time with no resting period, e.g. *Hypnum cupressiforme*; 2 a 10 month development with a short winter resting phase, e.g. *Mnium hornum*; 3 a 14–18 month development time with a long resting stage, e.g. *Plagiothecium sylvaticum* (Greene 1957). In *Mnium hornum* growing in the British Isles, the egg is fertilized in late spring and then secretes a wall around itself, grows and divides. Two growing points are quickly developed which respectively give rise to the foot and lower portion of the stalk and to the capsule and upper portion of the stalk. During development, the sporophyte is covered by the remains of archegonial tissue which increases in size to become the calyptra. The calyptra helps to protect the stalk and capsule during growth and over the winter resting phase. It also has other important functions mentioned below. In *Mnium* the capsule stalk elongates and the calyptra is shed by early April. Meiosis of the spore mother cells occurs when the capsules are fully grown but still green and slightly translucent. The operculum falls off and spores are shed in the latter half of April. By midsummer, the old capsules have become detached and are no longer found. Most of the research on the function of the different parts of moss sporophytes has been done on *Funaria hygrometrica* and a lesser amount on *Polytrichum*.

2 FUNCTION OF THE DIFFERENT PARTS OF THE SPOROPHYTE AND OF THE CALYPTRA

2.1 The foot

The foot becomes embedded in gametophore tissue and its principle function is absorbing the water, inorganic ions and nutrients required for the growth of the remainder of the sporophyte. It also has a role as an attachment and anchoring organ late in development when the capsule stalk elongates and the capsule expands. At the base of the foot where gametophyte and sporophyte tissue meet, transfer cells are developed on both. These are cells which are specialized for the efficient transfer of nutrients by the differentiation of a cell wall labyrinth and formation of dense cell membranes and numerous organelles. The labyrinth is formed by growth of cell wall material into the protoplasts which increases its surface (Browning & Gunning 1977) area much as the villi of animal gut increases the absorption area in those organisms.

Water moves up the centre of the leafy moss gametophore to the centre of the base of the foot. Here it crosses by osmosis from the transfer cells of the gametophyte tissue to those of the sporophyte tissue. It is then conducted up to the developing capsule through the hydroid elements of the stalk. Around the central region of the foot base is a zone where the transfer cells are best developed and enzymatically very active. This is the area where photosynthate and other nutrients move into the foot from the gametophore. As the capsule matures, the cell wall labyrinth fills up with new wall material and there is a corresponding loss of ability to transfer nutrients. This correlates with the change of function from an absorbing to an anchoring organ (Hébant 1975; Wiencke & Schulz 1978).

2.2 The stalk

Before the young sporophyte is 8 mm long, the apical region becomes destined to form the upper part of the capsule (the spore dispersal apparatus, the spore sac and some photosynthetic tissue). The subapical region will give rise to the lower part of the capsule (photosynthetic tissue and stomata) and also a meristem which divides and is thus responsible for elongation of the capsule stalk (French & Paolillo 1975a) The stalk will cease to grow after two days if the apical region of the young sporophyte is removed. However this growth is largely restored if the plant hormone cytokinin (in the form of benzyl adenine) is applied to the cut end. The same hormone enhances stalk growth and suppresses capsule expansion in non-decapitated young sporophytes. To date it is not clear whether the apical region produces cytokinins or maintains physiological conditions in the subapical region which allows synthesis of the hormone (French & Paolillo 1975b).

Water, inorganic ions and nutrients are transferred from the gametophore to the sporophyte via the stalk which possesses a well developed conducting tissue for this purpose (Hébant 1975). In *Mnium hornum*, at the stage when capsule expansion is just beginning, it has been shown that radioactive carbon dioxide fixed by the leaves on the gametophore, moves into the capsule stalk within 24 hours and then accumulates in the young capsule. Experiments on fully grown capsules, which can fix considerable amounts of radioactive carbon dioxide, have shown no evidence for photosynthate movement in the reverse direction, i.e. from the sporophyte to the gametophore (Proctor 1977).

2.3 The calyptra

This is formed from gametophyte tissue and is the membranous cap which covers the capsule until nearly mature. The calyptra protects the developing sporophyte and its removal leads to abnormality. Thus if the calyptra is taken off before the sporophyte is 10 mm long, there is permanent postponement of capsule development. Once the sporophyte has developed much beyond this length, expansion of the capsule will occur although the stalk will be abnormally thick. Early removal of the calyptra at a time which still allows capsule development, results in stomata lacking their usual orientation to the long axis of the capsule. Removal later, at about the time when the guard mother cells divide, leads to abnormally shaped guard cells (French & Paolillo 1975c). Replacing calyptrae with others which have been extracted with acetone and ethanol, results in normal development of stalk, capsules and stomata. This is clear evidence that the calyptra acts by physically containing the developing capsule and not by controlling sporophyte development through hormone secretion. However the presence of the calyptra seems to be essential to provide an environment facilitating the hormonal change in the young developing capsule (French & Paolillo 1975d).

2.4 The capsule

The structure of the capsule as it relates to spore dispersal will be described subsequently. Basically a moss capsule is made up of a sterile photosynthetic lower portion (the apophysis), a spore sac and a dispersal mechanism. One important question is the extent to which photosynthesis by the lower portion of the capsule provides the necessary sugars for growth and spore production. As early as 1886, it was shown that capsules cut off just as the spore mother cells were being differentiated would continue to grow on inorganic media. Capsules of *Funaria hygrometrica* increased their dry weight by 150% in three weeks and produced normal spores. More recently it has been estimated that the capsule of this moss may be responsible for about 50% of its own nutrition during the latter part of capsule expansion when both the rate of photosynthesis in the capsule and translocation from the gametophore reaches a maximum (Proctor 1977). The importance of light for capsule development in this species is further underlined by the observation that the dry weight of capsules attached to the gametophore or explanted onto sugar-containing agar increases with the intensity of illumination. In complete darkness morphology is abnormal and the spore sac fails to develop. Light intensities above 2 W/m^2 (0·6 k lux) enable development of normal capsules, but near the lower limit, capsules of reduced weight will form (French & Paolillo 1976).

The situation in *Polytrichum* is rather different as the capsules seem unable to carry out sufficient photosynthesis to balance respiration over a 24 hour period (Paolillo & Bazzaz 1968). The dependence of the sporophyte on the gametophore in this genus was confirmed in experiments where the gametophore was defoliated or the sporophyte was covered to exclude light (Grubb 1970). The paradox is that, among the mosses, *Polytrichum* has one of the more complex capsules which might be supposed best able to make a contribution towards growth of the capsule. It is probable that the synthetic flexibility provided by having a functional photosynthetic system and the ability to fix a considerable part of the carbon dioxide respired during growth and spore formation is more important than self sufficiency. In terms of carbon assimilation, Proctor (1977) has pointed out that it is best to regard a moss gametophore and sporophyte as an integrated whole. A parallel situation is seen in higher plants where, for example, only 25% of the total carbon requirement of the developing seeds in a pea pod is provided by photosynthesis of the pod itself. In most other mosses the contribution to the growth of the capsule, provided by photosynthesis in the capsule, probably lies between the two examples just described. The capsules of ephemeral mosses seldom have stomata or much chlorophyll and so contribute even less than *Polytrichum*.

Active photosynthesis in moss capsules is facilitated by the presence of stomata in the capsule wall. Two types of stomata are found depending on the species, round-pored and long-pored: the latter are more common. The capsules of most mosses possess less than 20 stomata but genera having 100 or more stomata include *Bryum*, *Philonotis*, *Physcomitrium*, *Plagiobryum* and *Pohlia*. The exact number is frequently species specific. Thus in *Funaria hygrometrica* there are up to 200 stomata while other species of the same genus have less than 100. The capsules of *Polytrichum* contain large long-pored stomata which vary from 20 (*P. alpestre*) to nearly 200 (*P. formosum*). At the other extreme *Rhacomitrium* capsules have up to 30 small stomata. Quite a number of mosses, e.g. *Leucobryum*, have no stomata on the capsule while in *Sphagnum* they are non functional. In the latter group some diffusion of gases through the capsule wall presumably occurs (Paton & Pearce 1957).

Although early studies suggested that the stomata on moss capsules only close under drought conditions (Paton & Pearce 1957), it is now clear that at least in *Funaria hygrometrica*, they also readily respond to light from the 5th to the 10th day of capsule expansion. In this species, closure occurs in the dark or following treatment with the plant hormone abscisic acid. The ability to respond declines as the capsule matures so that in the late stages only about 50% of the stomata are closed at night (Garner & Paolillo 1973).

3 SPORE DISPERSAL MECHANISMS

There is a wide range of different mechanisms to assure dispersal of spores from moss capsules. Space limits the examples that can be described to a few of the more interesting. The number of spores dispersed from a single capsule varies from 4–28 large spores (200 μm in diameter) in *Archidium alternifolium*, a moss that grows on wet meadows and elsewhere, to more than 50 million (8 μm in diameter) in *Dawsonia* which grows in the New Zealand forests (Bland 1971; Kreulen 1972; Puri 1973).

3.1 Andraeaopsida

The capsules of granite mosses are elevated by the growth of the gametophore apex which forms an organ termed a pseudopodium, the seta remaining very short. The capsule is unusual in not containing photosynthetic tissue although like other mosses it is protected by a calyptra. This falls off at maturity and the capsule opens along longitudinal lines

Fig. 5.1 Capsules of *Andreaea rothii* which disperse the contained spores by splitting into four valves. This moss is characteristic of hard siliceous rocks in montane and arctic regions (photo: M.C.F. Proctor).

of weakness (normally 4 in number). These give rise to slits which do not extend to the apex or base of the capsule so that on opening four valves are formed (Braithwaite 1887; Fig. 5.1). During dry weather the valves open so that the spores can be dispersed and close during moist conditions (Parihar 1977).

3.2 Sphagnopsida

Capsules of peat mosses are red to black brown, spherical and 2–3 mm in diameter when mature. As in the granite mosses they are elevated by pseudopodium which may reach 3 cm in length. At the top of the *Sphagnum* capsule is a disc shaped lid or operculum which is separated from the rest of the capsule by a circular groove, the annulus. Spore discharge in peat mosses is readily observed if ripe capsules are brought into a room and placed a few metres from a fire. Under natural conditions discharge occurs on dry sunny days (Ingold 1965). A section

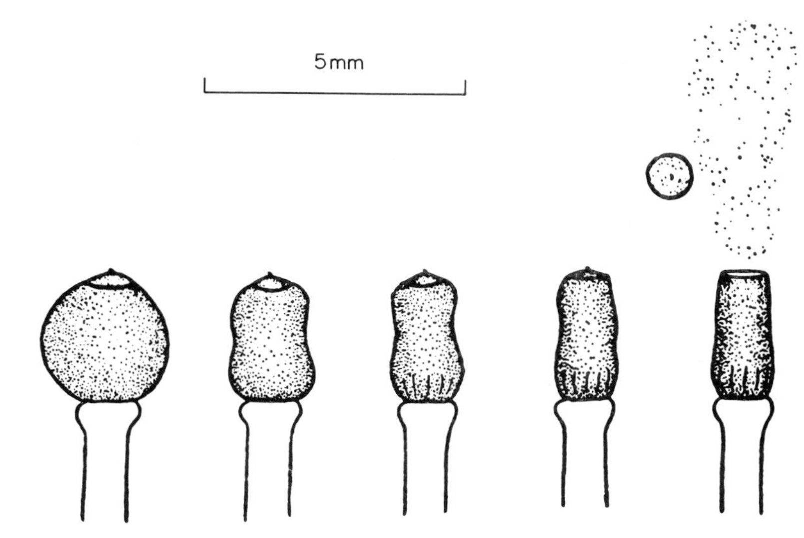

Fig. 5.2 Stages in the drying of *Sphagnum* capsules leading finally to discharge of the spores (Ingold 1965).

through a nearly mature capsule shows a dome shaped spore sac and parenchymatous columella. The latter transfers nutrients to the developing spore sac and then breaks down as the capsule matures. Air can enter the capsule, probably by way of the thin walled guard cells of the non-functioning stomata while the enlarging capsule is still moist. This results in the large cavity left by the degenerating columella becoming filled with air. Once the wall of the capsule dries it is impermeable to gases and the thickening of the wall cells is such that they contract transversely but not longitudinally as they lose water. Thus the capsule becomes cylindrical as it dries and the air within compressed to 4–6 atmospheres (Fig. 5.2). As the cells of the operculum do not shrink, stresses are set up in the area of the junction of the lid and the capsule. Finally the combined effect of stress and internal pressure results in the operculum being thrown off and the spores shot several centimeters into the air. The noise, a click, which accompanies the discharge can be heard from several metres and the whole process is colloquially known as a 'pop-gun mechanism' (Maier 1974).

Sphagnum capsules are elevated at maturity by the growth of the pseudopodium so that they are above the water level in the mires where most peat mosses grow. However, capsules may sometimes be submerged by changing water levels and may then discharge their spores by a different mechanism. The operculum is not shed. Instead the capsule falls off the pseudopodium, the base of the columella decays and the spores are liberated through a hole in the capsule base. Alternatively the spores may germinate while still enclosed in the capsule and burst it by expansion. When this happens, both the lid and base may be forced off leaving the barrel-shaped shells which are commonly found in *Sphagnum* tufts (Braithwaite 1880).

3.3 Bryopsida

In most true mosses there is a lid or operculum, at the apex of the capsule, and below this one or two whorls of teeth (the outer and inner peristome) whose movements determine the pattern of spore liberation (Edwards 1980). Where the lid is attached to the capsule there is a row of special cells, the annulus. Under natural conditions, in the presence of liquid water (dew or rain), the operculum is thrown off so rapidly that direct observation of the process is almost impossible. In the laboratory it has been shown that the cells of the annulus absorb water and 'unzip'. As they do this stresses are set up in the capsule, which becomes deformed, so providing the potential energy which throws the cap off as the capsule regains its proper shape (Maier 1973). The following examples give some idea of the variety of spore dispersal mechanisms in the true mosses. About half of all mosses have single peristomes, the other half double, with a different origin and a different mode of operation (See Edwards 1979).

(a) *Capsules without peristomes*

These are found most commonly in ephemeral mosses which grow in habitats that are only temporarily available, for example, bare spots in fallow fields or areas of natural soil erosion. Such capsules almost always have mouths that point vertically. In the aptly named moss *Ephemerum*, sometimes called 'emerald dewdrops', the capsule stalk remains very short so that the capsule is cradled in a cluster of leaves at the apex of the tiny gametophore. Stomata are seldom present on the capsule wall and there is no operculum or annulus so that the large spores are dispersed by irregular splitting of the capsule (Fig. 5.3). *Physcomitrium* is another ephemeral moss but is slightly more robust and most frequently seen on mud at the edges of ponds and heavy

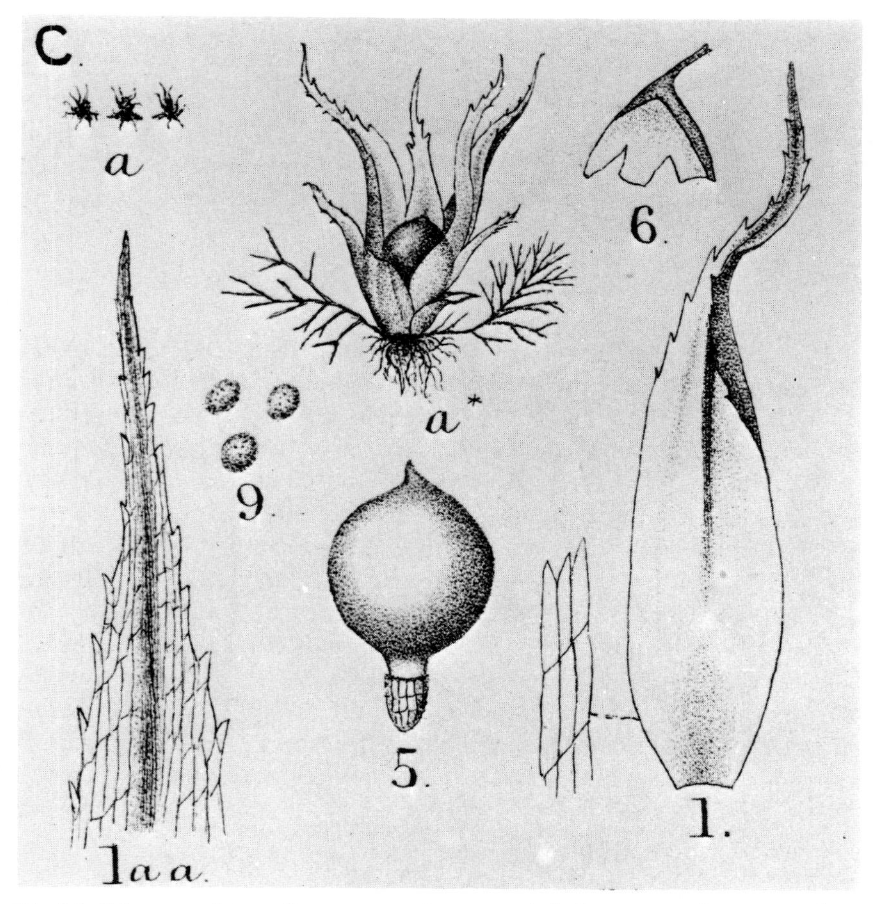

Fig. 5.3 *Ephemerum intermedium*, a moss whose capsule lacks an operculum or annulus (a) whole plant, (1) leaf, (5) capsule, (6) calyptra (9) spore (Braithwaite 1887).

calcareous soils. The capsule has an operculum with an annulus and contains some 10,000 moderately large spores while the stalk is again very short. On drying, the cells below the wide mouth of the capsule constrict which presumably helps to throw off the operculum and leads to this moss being called the urn moss (Fig. 5.4). The cup-shaped capsules of these mosses allow spores to be sucked out by the gentle vortex which forms as air blows across the top. Ephemeral mosses are usually minute in size and have a telescoped life history so that they are only seen in early spring or late autumn. In spite of the large size of the spores and seemingly inefficient dispersal mechanisms, ephemeral mosses are widely distributed (Crum 1972).

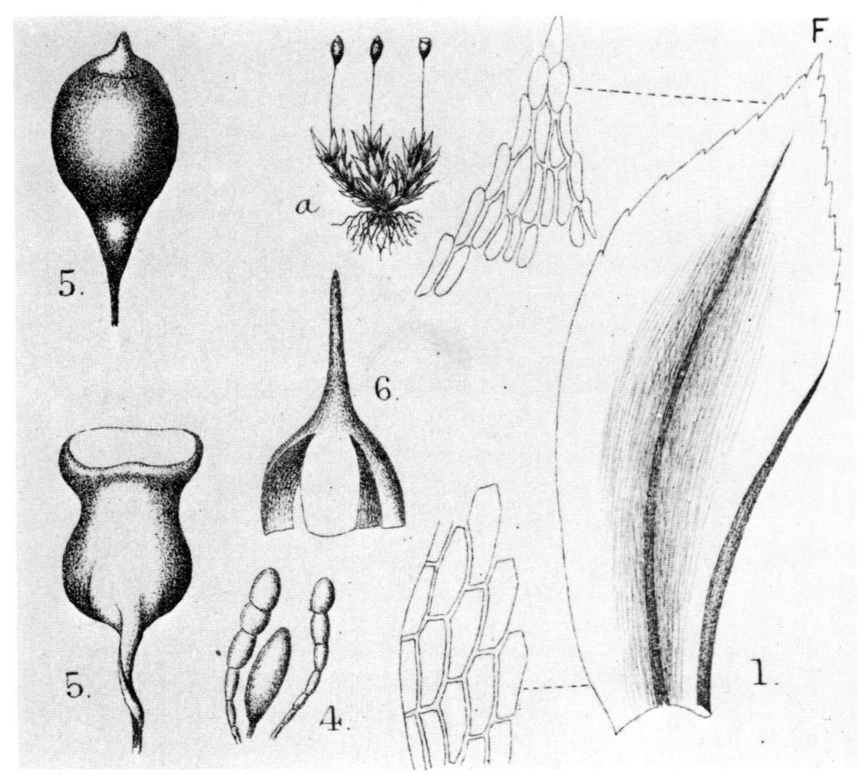

Fig. 5.4 *Physcomitrium pyriforme*. Spore dispersal is aided by a vortex produced by air currents. (a) whole plant (1) leaf, (4) antheridia and paraphyses, (5) capsule before and after loss of operculum, (6) calyptra (Braithwaite 1887).

(b) *Capsules with single or poorly developed peristomes*

These mosses also have erect rather than inclined or pendant capsules. Two examples suffice. *Tetraphis pellucida* grows on stumps and peat. The capsule is endowed with four massive peristome teeth which close over the apex under wet conditions (Fig. 5.5). In contrast to this simple dispersal mechanism, this moss has developed a very advanced means of vegetative reproduction. Cup-like endings are found on the tops of the shoots of the gametophore, which are filled with minute green gemmae, so that survival of the species seldom depends on successful production and dispersal of spores (See Fig. 6.7–26 & 27). In *Buxbaumia*, a peculiar peristome membrane is seen when the operculum

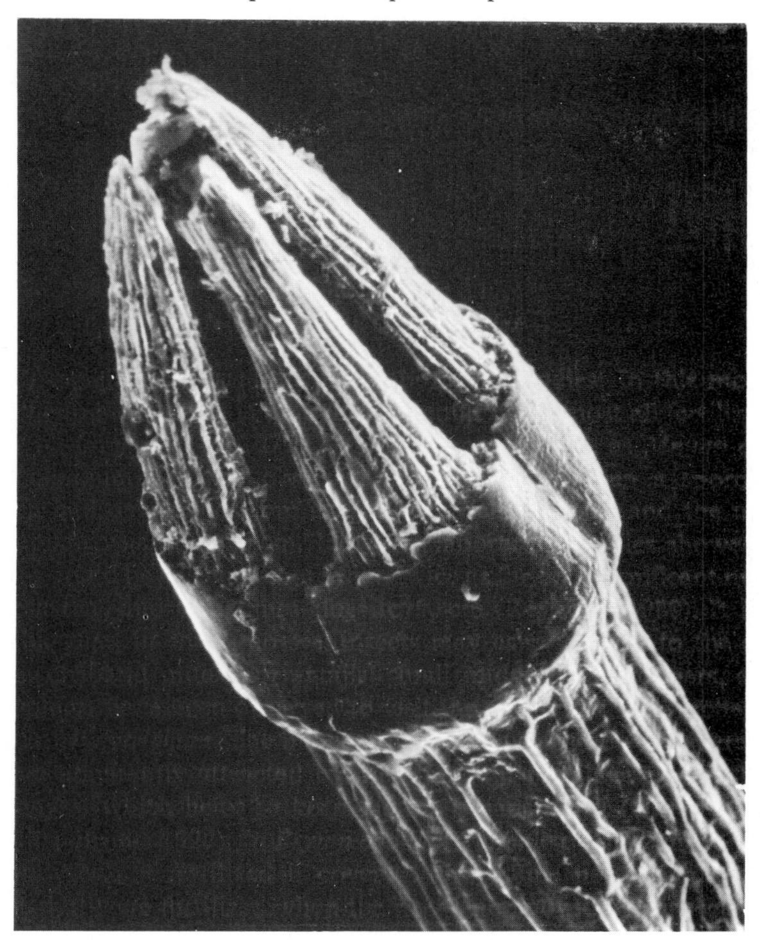

Fig. 5.5 A scanning electron micrograph of the top of the capsule of *Tetraphis pellucida*. The capsule is unique in having only four large peristome teeth. The chief habitat of this moss is rotting tree stumps (photo: Biological Sciences Department, Exeter University).

is thrown off from the remarkably large capsule (Fig. 5.6). It is the size of a *Polytrichum* capsule and elegantly tear-drop shaped containing about 5 million spores (Kreulen 1972). The capsule is quite oversized in proportion to the stalk (up to 2 cm long) and to the gametophore which forms a blackish crust, composed partly of protonema, on burned soil. Rain pelting on the broad upper surface of the capsule causes spores to be emitted in little puffs (Crum 1972). This moss has been suspected of being saprophytic because it is difficult to imagine how such a small

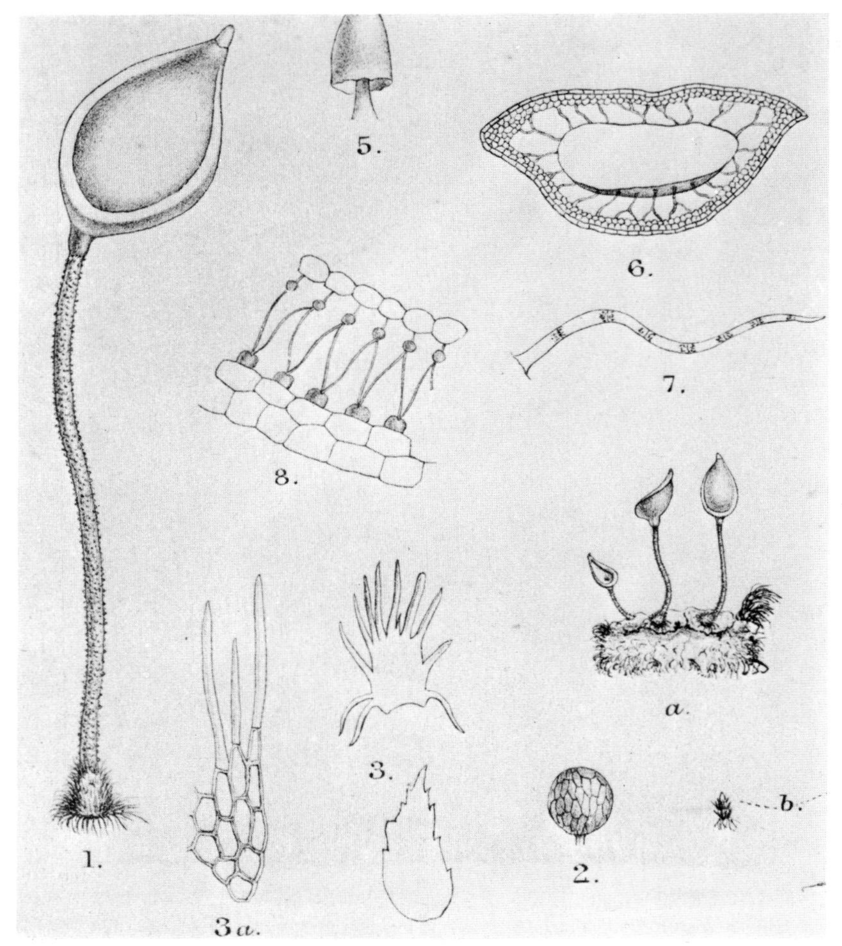

Fig. 5.6 *Buxbaumia aphylla* dispersal of the large spores is aided by rain drops falling on the capsule. (b, a) whole plant, (1) the capsule, (2) antheridia, (3) perichaetial leaf (5) operculum and part of columella, (6) section through the middle of the capsule, (7) filaments connecting the spore sac with the capsule wall, (8) transverse section through operculum and peristome (Braithwaite 1887).

amount of photosynthetic tissue in the tiny gametophore and protonema could provide sufficient nutrients for the huge capsule. However, the capsule contains abundant chlorophyll and has at least 20 stomata (Paton & Pearce 1957) so that the matter requires verification. The capsules of this species are also unusual in being phototropic. Although initially more or less erect, they become tilted so that the pointed end is directed towards the greatest incident light, i.e. usually south or south-west in the N. hemisphere (Crum 1973).

(c) *Capsules with double peristomes*

In mosses with two rows of teeth, the peristome is formed from a single layer of cells which forms a dome beneath the operculum of the capsule. This layer of cells becomes thickened on both the internal and external walls but the lateral walls and part of the transverse wall remain unaltered. Eventually the outer and inner walls break apart and each becomes split longitudinally to form the peristome teeth observed in the mature capsule. Usually, only the outer teeth are hygroscopic and this results from their being thickened on their outer side with material that absorbs water. The outer side of the tooth lengthens as water is absorbed while the inner side is not affected by changes in water content. Thus the outer peristome teeth are essentially two layered structures. Under moist conditions the outer side of the outer peristome increases in length and the teeth bend inwards to reform a dome over the apex of the capsule preventing further spore discharge. As the teeth dry the outer side of each tooth shortens bending backwards with jerky movements as it separates from its neighbours. The inner teeth act as a sieve allowing gradual discharge of the spores a few at a time (Edwards 1979). While it is often stated that the teeth bend outwards under dry conditions to allow spore dispersal this is an oversimplification. Such movement occurs in epiphytic mosses with reduced peristomes (e.g. *Orthotrichum*) but in other mosses the teeth often bend inward under wet or completely dry conditions. Thus dispersal of the spores from the capsule takes place just as rain begins or shortly after it ceases. At this time the spores would presumably have the greatest chance of germinating successfully wherever they finally come to rest (Lazarenko 1957). The peristomes of a few genera are even reported to open under wet conditions (Patterson 1953).

Funaria hygrometrica. Besides being a popular subject for experimental investigations, this common moss is one of the first colonizers of ground following fire. The inclined capsules contain up to about 500,000 spores. The peristome of this moss is unusual in that the outer teeth are joined at their tips to a small central plate of cells (Fig. 5.7). Under damp conditions these teeth absorb water and elongate but because they are fixed at both apex and base, the elongation has the

Fig. 5.7 A scanning electron micrograph of the top of a capsule of *Funaria hygrometrica*. The outer red teeth are twisted spirally to the left and converge at the centre where they are joined to a small central disc. This moss is found where recent fires have occurred in fields, woods and gardens (photo: Department of Biological Sciences, Exeter University).

effect of increasing the curvature of the teeth. This results in the whole peristome undergoing a twisting movement that brings about closure of the slits between the teeth. Under dry conditions movement in the reverse direction provides space between the inner and outer teeth which allows the spores to be shed (Ingold 1965). The long delicate capsule stalk also aids the process as it too shows hygroscopic movements. Thus, during changes in humidity, the stalk twists or untwists (under moist conditions) so that the capsule rotates quite rapidly. The centrifugal force and jerky movements of entangled capsules enhances dispersal of the spores.

Eurhynchium confertum. This is a common moss of garden walls, stones and tree bases. It illustrates the point that, in addition to the salt shaker type of dispersal, moss peristomes have the potential for a forcible type of spore discharge. The outer peristome teeth in this species are quite free and bend outwards under dry conditions so that the spores can sift out between the stiff inner teeth. In the gap between these inner teeth there are also 2 or 3 flexible hairs (cilia). Under damp conditions, the outer peristome teeth bend rapidly inwards and the tips poke through the gaps between the inner teeth depressing the cilia as they do so and becoming, in the process, covered with spores. When drier conditions return, the outer teeth try to bend back to their former position but are momentarily restrained by the stiff inner teeth so that a rapid jerky movement ensues which flings the spores out several millimetres. About 100 spores may be dispersed by each backward movement of the teeth and if repeated many times up to about 5% of the spores (16,000 of 300,000) in a capsule might be forcibly discharged (Ingold 1965; Kreulen 1972).

Timmia bararica. In this moss from the U.S.S.R. and central Europe, active dispersal is even more effective. Here the 16 teeth of the outer peristome are joined in pairs at their tips to form 8 triangles (Fig. 5.8). Each pair moves hygroscopically squatting down and bulging out in the middle part while the tips bend in to plough the spores from between the inner teeth and the spores are thrown out rather than merely falling (Lazarenko 1957).

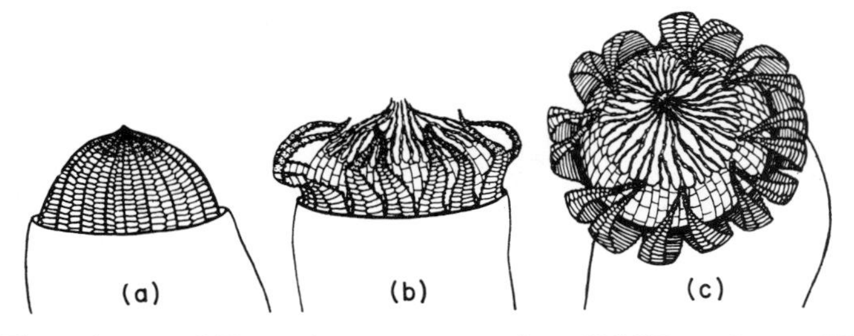

Fig. 5.8 The peristome of *Timmia bavarica*, a moss from U.S.S.R. and central Europe in which the peristome teeth actively help to disperse the spores (a) when wet, (b) when dry, (c) in motion (Lazarenko 1957).

Other examples. A number of mosses possess peristome teeth which are very long. These may be twisted remarkably as in *Tortula, Barbula* and some species of *Tortella.* In moist weather, the twisted bristle-like teeth do not allow the spores to escape but under dry conditions the teeth separate sufficiently for the spores to be dispersed (Fig. 5.9). In aquatic mosses, capsules generally form when water levels are low and the plants exposed. *Fontinalis* has a well developed bright red

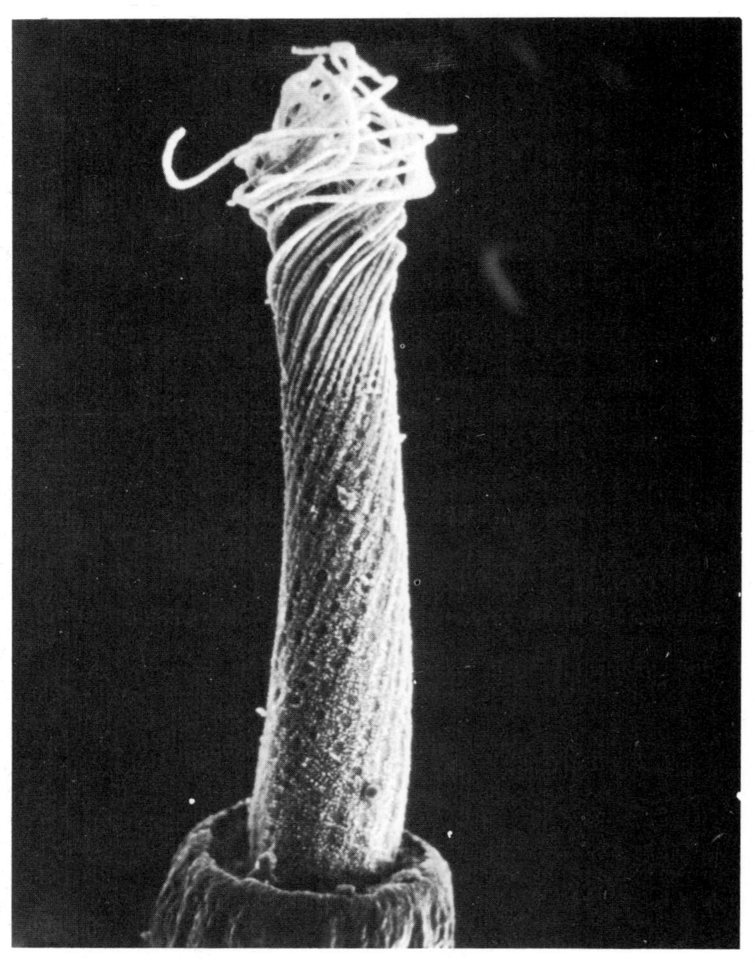

Fig. 5.9 A scanning electron micrograph of the apex of the capsule of *Tortula muralis.* The 32 long twisted teeth are united at the base and in damp weather close over the mouth of the capsule. In dry weather they separate and allow dispersal of the spores. The moss is found on brick and stone walls (photo: Department of Biological Sciences, Exeter University).

peristome. The inner peristome consists of 16 teeth which form a delicate network with square perforations. These are thought to prevent the spores being washed out too quickly (Puri 1973).

(d) *Capsules with teeth and an epiphragm*

The Polytrichales includes mosses such as *Polytrichum* and *Atrichum*.

Fig. 5.10 A scanning electron micrograph of the apex of the capsule of *Atrichum undulatum*, a moss closely related to *Polytrichum*. The spores are dispersed by a 'pepper pot mechanism'. *Atrichum* is moss which indicates a good humus content in the soil and clearings in woods are a favourite habitat (photo: Department of Biological Sciences, University of Exeter).

The capsule of *Polytrichum* contains up to 2 million spores whose dispersal is controlled by a single whorl of 32 short peristome teeth which are attached by their tips to a membranous tissue, the epiphram. This stretches drum-like across the top of the capsule. Under dry conditions, the thinner-walled cells between the teeth lose water and the spores sift out through slit-like gaps between the teeth (Fig. 5.10). Wind tunnel experiments show that almost no spores fall from the capsule if it is held still in an air stream (Ingold 1965). However if the capsule is free, wind, rain drops or the movement of animals, cause the stalk to vibrate

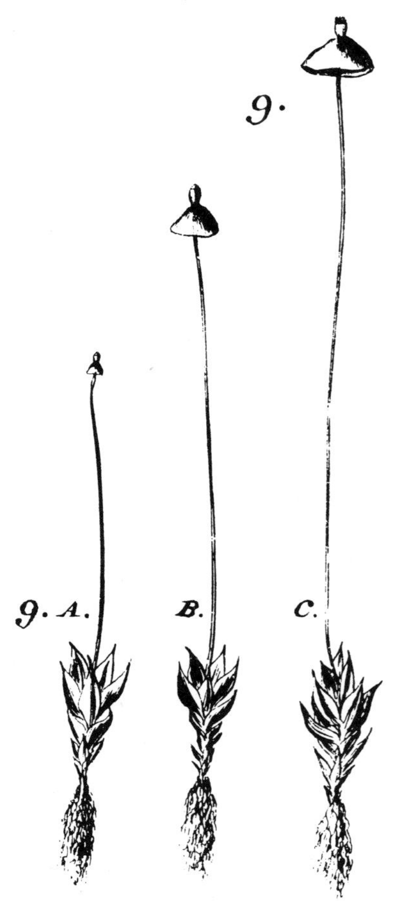

Fig. 5.11 Capsules of *Splachnum* sp. as originally drawn by Dillen (1741) who correctly observed that the stalk of this moss, unlike others, continues to grow even after the capsule is mature.

rapidly, the high amplitude being due to the great length and elasticity of the dry wiry stalk (Sarafis 1971). The result is that spores are shed in large numbers.

Rain is also involved in spore dispersal from *Dawsonia*, a relative of *Polytrichum*. The capsules are erect at first but at maturity become nearly horizontal. Raindrops falling on the upper surface of the capsule depress it and spores are puffed out between its long twisted teeth (Goebel 1906).

(e) *Capsules with insect-dispersed spores*

No account of spore dispersal in mosses could omit some description of the Splachnaceae. They were first described in 1695 and illustrated shortly afterwards by Dillen in 1741 (Fig. 5.11). These mosses grow on decaying dung or carrion (Table 5.1), particularly that dropped on *Sphagnum* bogs or among *Sphagnum* plants on boggy mountain sides. Presumably this moist habitat ensures that the dung does not dry out or decay too quickly. *Splachnum* and its relatives are mostly found in arctic and boreal regions although they also occur in the tropics at high altitude.

Table 5.1 The variety of dung and carcass colonized by some members of the Splachnaceae

Moss	Colour of apophysis	Dung colonized
Haplodon wormskjolffi	Brown-purple	Human, carcass of sheep, lemmings
Splachnum ampullaceum	Light purple	Reindeer, moose, cattle, sheep
S. luteum	Yellow	Reindeer, moose
S. rubrum	Dark red	Reindeer, moose
S. ovatum	Dark purple	Cattle
S. vasculosum	Dark purple	Sheep, reindeer, decomposing whale meat
Tayloria longicollis	Brownish green	Human
Tetraplodon ongustatus	Light brownish	Carnivores and animal carcass
T. mnioides	Purplish black	Sheep, reindeer, decomposing sheep, yak

Data from Bequaert (1921), Crum (1972), Iwatsuki & Steere (1975), Koponen & Koponen (1978), Nyholm (1954), Steere (1973), Watson (1955).

Splachnum spp. form rather dense cushions up to 25 cm across and the capsules develop upon a thick translucent stalk that continues to grow even when the spores are ripe (Steere 1973). The stalk may in *S. luteum* reach 20 cm in length. The neck of the capsule develops a swollen region, the apophysis, up to 2 cm across in some species. At maturity, the operculum falls off to reveal 16 teeth which first bend back

Fig. 5.12 A scanning electron micrograph of the top of the capsule of *Splachnum ampullaceum* showing the peristome teeth and columella. Note the sticky spores attached to the latter. The capsule has a greatly extended lower area (Fig. 5.11) which emits odours that attract flies (photo: Department of Biological Sciences, Exeter University).

and then become reflexed due to shrinkage of the capsule wall. This also results in the columella projecting beyond the rim so that the mass of spores is pushed up and some fall on the outside of the capsule while others are trapped among the peristome teeth (Fig. 5.12). The spores of these mosses are quite unlike those of other mosses in being sticky. Shaking will not dislodge them from the capsule. In moist weather the capsule swells to its former size which takes the spore mass on top of the columella back into the capsule while the teeth fold back and close the mouth (Bequaert 1921; Ingold 1965).

The apophysis, sometimes skirt-shaped, becomes brightly coloured at about the time the spores are ripe. Its colour can be purplish (*S. ampullaceum*), red (*S. rubrum*), yellow (*S. luteum*) or dazzling white (*Tayloria mirabilis*) (Koponen 1978) so that a capsule resembles an entomophilous flower though at a distance a cluster often looks somewhat like a large coloured toadstool. A detailed examination of the apophysis reveals about 50 stomata which are the largest found in mosses (Bequaert 1921; Paton & Pearce 1957). The stomata are reputed to emit the various odours that are characteristic of mosses in this family. At least in the case of *Tetraplodon mniodes*, which has a stout red stalk and purplish-black capsules, they also seem to exude a secretion which is 'licked up' by flies. These insects, attracted by smell and secretion, inevitably become coated with a number of the sticky moss spores when alighting on, or leaving the capsule. It has been observed that flies that have visited a *Splachnum* patch often subsequently go in search of fresh dung, thereby disseminating the moss spores to a suitable substrate for their development.

Recent studies show how attractive *Splachnum* species can be to coprophilous flies. Eight traps baited with *S. vasculosum*, *S. ampullaceum* or *S. luteum* caught 135 insects compared with 18 in control traps. Of the insects trapped (some fortuitously) by the *Splachnum* baited traps, 43 were of a kind unlikely to carry the spores to fresh dung. However the remainder included Scatophagidae which lay their eggs on dung, coprophilic Muscid flies, and flies of the Pyrellia–Orthellia group that may in the past have been mistakenly called blow-flies. The latter may feed on dung although they rarely breed on it (Koponen & Koponen 1978).

Several questions remain in connection with these beautiful and remarkable mosses. What is the nature of the odoriferous substance and secretions and do they vary between species so that particular flies are attracted to particular species of moss (Koponen & Koponen 1978)?

Preliminary results indicate that the odour of *Splachnum luteum* is due mainly to 1-octen-3-ol, a compound with a strong mushroom-like smell. Lesser amounts of 3-octanone, 3-octanol, octanol and 2-octen-1-ol are also present. In *Splachnum vasculosum*, *Tetraplodon mnioides* and *Tayloria tenuis*, there is a small amount of 1-octen-3-ol but the main volatile compound is butyric acid which, to man, has an unpleasant odour. Acetic acid and proprionic acid are also synthesized by these species (Pyysalo *et al.* 1978).

4 EFFECTIVENESS OF THE DISPERSAL MECHANISMS

In general, mosses produce vast numbers of spores e.g. in Canada *Pleurozium schreberi* fruits abundantly and a carpet of this moss may produce 100 million healthy spores per square metre (Longton 1976) which are usually 8–12 μm in diameter. These may be carried as much as 19,000 k, given a combination of fortuitous circumstances, especially in the climatic zones controlled by cyclonic wind systems. There is some controversy as to whether such long distance dispersal of moss spores results in the successful establishment of new moss colonies. Positive evidence comes from a recent study in New Zealand which showed a correlation between the resistance of spores to drought and freezing and the geographical areas colonized by particular moss species (Zanten 1978). The other view is that while mosses are somewhat more diffusely distributed than seed plants of similar range, long distance dispersal is not a very important means for the dispersal of most mosses. It is pointed out that rather demanding ecological conditions are frequently required by the germinating spore on landing for it to be able to establish a new moss colony successfully. In addition, in dioecious mosses, at least two long-distance dispersed spores of the same species need to land in the same vicinity (Crum 1972).

The spores of some mosses can certainly withstand the harsh conditions expected during long-distance dispersal. Thus spores of *Ceratodon purpureus* and *Funaria hygrometrica* can germinate after drying for 16 and 13 years respectively and these two are cosmopolitan weeds. However the ephemeral moss. *Physcomitrium turbinatum* is also very widely distributed and yet has large spores, an apparently inefficient dispersal mechanism, and spores which fail to germinate after drying for more than a few months. Thus for this moss, long-distance dispersal would seem unlikely. As an alternative, the present distribution of mosses may be explained by assuming that dispersal over short distances (metres to tens of kilometres) is the norm; that they have a slow rate of speciation; and that they move along the natural migratory routes along with higher plants (Longton 1976). *Pleurozium schreberi* is an extremely common moss in the boreal forest zones of the northern hemisphere and is also recorded from every vice county in the British Isles. However, in Europe, this moss rarely produces capsules and has no special means of vegetative reproduction. Abundant spore production and hence the possibility of long-distance dispersal is not therefore a prerequisite for wide distribution (Crum 1972). Another instance where long distance dispersal is very unlikely is in the Splachnaceae whose spores are carried by insects and often require dung of particular animals for growth. It is thought that these unusual mosses must have spread to the various continents along the land connections before India and Australia separated from Africa some 200 million years ago (Koponen 1978). Earlier evolution of the taxa in this family is considered unlikely as their biology requires the presence of both coprophilous flies and large animals. Thus while the effectiveness of short distance dispersal is attested to by empirical observation of for example the colonization of burned sites (Chapter 9), and the abundance of moss spores in the air flora, the question of the importance of long distance dispersal requires further research. The present evidence indicates that successful long distance moss spore dispersal is rarely accomplished though the establishment of bryophyte floras on recent oceanic islands suggests that it is possible (Longton 1976).

5 REFERENCES

Bauer L. (1963) On the physiology of sporangium differentiation in mosses. J. *Linn. Soc. Lond. Bot.* **58,** 343–59.

Bequaert J. (1921) On the dispersal by flies of the spores of certain mosses of the family Splachnaceae. *Bryologist*, **24,** 1–4.

Bland J. (1971) *Forests of Lilliput: the Realm of Mosses and Lichens.* Prentice Hall, Englewood Cliffs.

Braithwaite R. (1880) *The Sphagnaceae or Peat Mosses of Europe and North America.* David Bogue, London.

Braithwaite R. (1887) *The British Mosses*, Vols 1, 2 & 3. Reeve, London.

Browning G.A. & Gunning B.E.S. (1977) An ultrastructural and cytochemical study of the wall-membrane apparatus of transfer cells using freeze-substitution. *Protoplasma*, **93,** 7–26.

Crum H. (1972) The geographical origins of the mosses of North America's eastern deciduous forest. *J. Hattori Bot. Lab.* **35,** 269–98.

CRUM H. (1973) Mosses of the Great Lakes Forest. *Contrib. Univ. Mich. Herb.* **10,** 1–404.

DILLEN J.J. (Dillenius) (1741) Historia Muscorum, Oxonii.

EDWARDS S.R. (1979) Taxonomic implications of cell patterns in Haplolepidous moss peristomes. In *Bryophyte Systematics,* eds. Clarke G.S.C. & Duckett J.G. Systematics Association special volume, no. 14, pp. 317–46 Academic press, London.

FRENCH J.C. & PAOLILLO D.J. (1975a) Intercalary meristematic activity in the sporophyte of *Funaria* (Musci). *Am. J. Bot.* **62,** 86–96.

FRENCH J.C. & PAOLILLO D.J. (1975b) Effect of exogenously supplied growth regulators on intercalary meristematic activity and capsule expansion in *Funaria. Bryologist,* **78,** 431–7.

FRENCH J.C. & PAOLILLO D.J. (1975c) The effect of the calyptra on the plane of guard cell mother cell division in *Funaria* and *Physcomitrium* capsules. *Ann. Bot.* (*Lond.*), **39,** 233–6.

FRENCH J.C. & PAOLILLO D.J. (1975d) On the role of the calyptra in permitting expansion of capsules in the moss *Funaria. Bryologist,* **78,** 438–46.

FRENCH J.C. & PAOLILLO D.J. (1976) Effect of light and other factors on capsule expansion in *Funaria hygrometrica. Bryologist,* **79,** 457–65.

GARNER D.L.B. & PAOLILLO D.J. (1973) On the functioning of stomates in *Funaria. Bryologist,* **76,** 423–7.

GOEBEL K. (1906) Archegoniatenstudien. X. Beitrage zur kenntnis Australischer und Neuseelandisher Bryophyten. *Flora* (*Jena*), **196,** 1–202.

GREENE (1957) The maturation cycle, or the stages of development of gametangia and capsules in mosses. *Trans. Br. Bryol. Soc.* **3,** 736–45.

GRUBB P.J. (1970) In Report of the autumn meeting of the British Bryological Society, 1969. *Trans. Br. Bryol. Soc.* **6,** 216–18.

HÉBANT C. (1975) Organization of the conducting tissue-system in the sporophytes of *Dawsonia* and *Dendroligotrichum* (Polytrichales, Musci). *J. Hattori Bot. Lab.* **39,** 235–54.

INGOLD C.T. (1965) *Spore Liberation.* Clarendon Press, Oxford.

IWATSUKI Z. & STEERE W.C. (1975) Notes on Himalayan Splachnaceae. *J. Hattori Bot. Lab.* **39,** 345–61.

KOPONEN A. (1978) The peristome and spores in Splachnaceae and their evolutionary and systematic significance. *Bryophytorum Bibliotheca,* **13,** 535–67.

KOPONEN A. & KOPONEN T. (1978) Evidence of entomophily in Splachnaceae (Bryophyta). *Bryophytorum Bibliotheca,* **13,** 569–77.

KREULEN D.J.W. (1972) Spore output of moss capsules in relation to ontogeny of archesporial tissue. *J. Bryol.* **7,** 61–74.

LAZARENKO A.S. (1957) On some cases of singular behaviour of the moss peristome. *Bryologist,* **60,** 14–17.

LONGTON R.E. (1976) Reproductive biology and evolutionary potential in bryophytes. *J. Hattori Bot. Lab.* **41,** 205–23.

MAIER K. (1973) Dehiscence of the moss capsule. II. The annulus: analysis of its functional apparatus. *Oesterr. Bot.* **122,** 75–98.

MAIER K. (1974) Ruptur der Kapsewand bei *Sphagnum. Plant Syst. Evol.* **132,** 13–24.

NYHOLM E. (1954) Musci. In *Illustrated Moss Flora of Fennoscandia, Vol. II.* CWK Elerup, Lund.

PAOLILLO D.J. & BAZZAZ F.A. (1968) Photosynthesis in sporophytes of *Polytrichum* and *Funaria. Bryologist,* **71,** 3335–43.

PARIHAR N.S. (1977) Bryophyta. In *An Introduction to Embryophyta, Vol. 1,* 5th ed. Central Book Depot, Allahabad.

PATON J.A. & PEARCE J.V. (1957) The occurrence, structure and function of the stomata in British bryophytes. *Trans. Br. Bryol. Soc.* **3,** 228–59.

PATTERSON P.M. (1953) The aberrant behavior of the peristome teeth of certain mosses. *Bryologist*, **56,** 157–9.
PROCTOR M.C.F. (1977) Evidence on the carbon nutrition of moss sporophytes from $^{14}CO_2$ uptake and the subsequent movement of labelled assimilate. *J. Bryol.* **9,** 375–86.
PURI P. (1973) *Bryophytes, a Broad Perspective.* Atma Ram, Delhi.
PYYSALO H., KOPONEN A. & KOPONEN T. (1978) Studies on entomophily in the Splachnaceae (Musci). I. Volatile compounds in the sporophyte. *Ann. Bot. Fenn.* **15,** 293–6.
SARAFIS V. (1971) A biological account of *Polytrichum commune. N.Z. J. Bot.* **9,** 711–24.
STEERE W.C. (1973) Observations on the genus *Aplodon* (Musci: Splachnaceae). *Bryologist,* **76,** 347–55.
WATSON E.V. (1955) *British Mosses and Liverworts.* Cambridge University Press.
WIENCKE C. & SCHULZ D. (1978) The development of transfer cells in the haustorium of the *Funaria hygrometrica* sporophyte. *Bryophytorum Bibliotheca,* **13,** 147–67.
ZANTEN B.O. van (1978) Experimental studies on trans-oceanic longrange dispersal of moss spores in the Southern Hemisphere. *Bryophytorum Bibliotheca,* **13,** 715–33.

CHAPTER 6
SPORES AND PROTONEMATA

1 SPORES

1.1 SIZE AND SHAPE

In the moss capsule, four spores are formed within the spore mother cell following meiosis. In many species, the four spores during subsequent

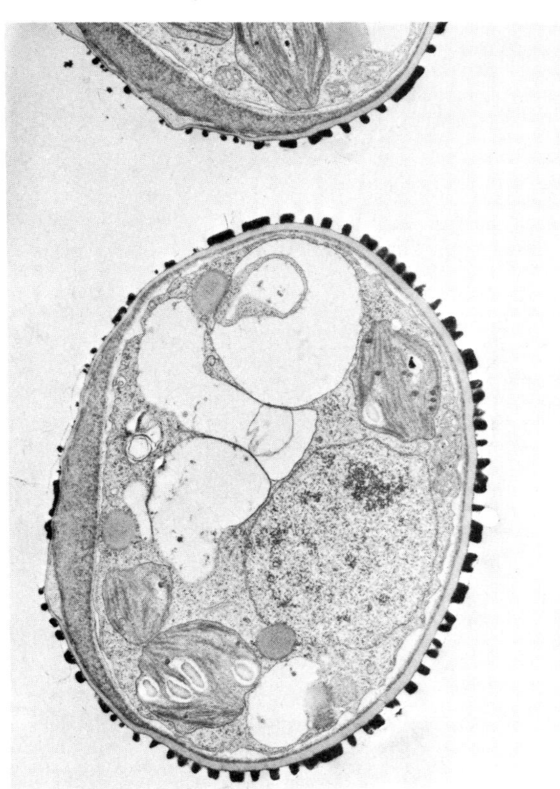

Fig. 6.1 A transmission electron micrograph of the spore of *Tetraplodon mnoides* which has been sectioned. Note the chloroplasts filled with starch grains, the vacuoles and large nucleus as well as the ornamentation of the outer wall. This moss colonizes decaying animal matter, often dead sheep (photo: Aune Koponen).

maturation compress one another and stick together so that even after later separation they have one large rounded side and three smaller flat triangular faces.

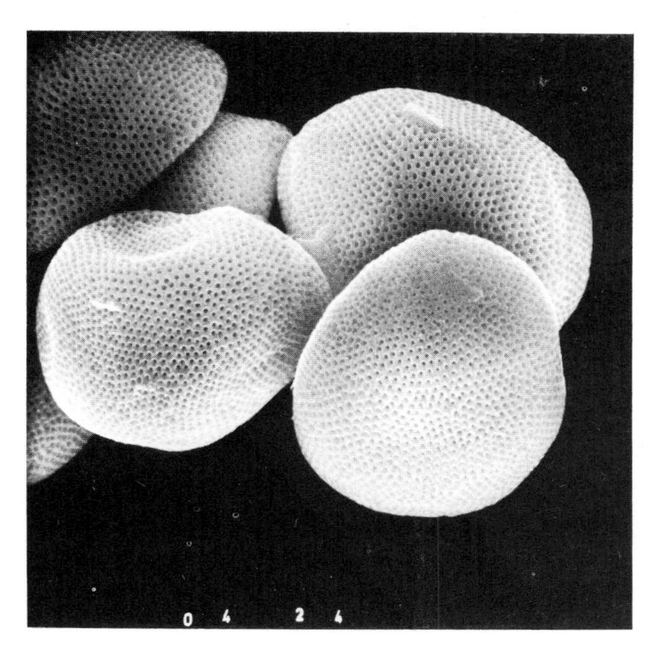

Fig. 6.2 A scanning electron micrograph of the spores of *Tetraplodon mnoides*. Note the ornamentation of the outer surface (photo: Aune Koponen).

The spores of granite mosses are about 30 μm in diameter and frequently remain united in four's when shed. The outer surface, the perine, is minutely warted and the interior filled with lipids and chloroplasts. Peat mosses have spores of about the same size, yellowish to brownish in colour, and subtriangular in shape; the perine may be smooth, wrinkled or granular depending on the species. The spores of true mosses are either green containing chloroplasts packed with starch grains or yellowish due to the presence of oil and in this case they lack green plastids. (Fig. 6.1) (Mueller 1974) and both size (5–200 μm) and sculpturing varies widely in different genera. The perine of many spores appears smooth when viewed with the light microscope although the electron microscope reveals greater complexity (Fig. 6.2). The spores of some genera such as *Orthotrichum* are highly sculptured (Boros & Jarai-Komlodi 1975).

1.2 GERMINATION

The moss spore wall is composed of three layers, the intine, the exine, and the perine. A localized area called the aperture region has been identified in electron microscope studies and it is here that water is absorbed. In the aperture region the intine is thicker and the exine thinner than over the rest of the spore. On being placed in water, a spore swells symmetrically at first as rehydration of the air dried cell occur. The first stage of germination follows in which uptake of water by the intine occurs. This results in asymmetrical swelling of the spore since this layer is much thicker in the aperture region. The exine is ruptured next, a process which takes from ten minutes in *Polytrichum commune* to many hours in *Funaria hygrometrica* (Heitz 1942; Olesen & Mogensen 1978). Respiratory activity is quickly observed but further germination will not take place in the dark except on artificial media containing organic substances, e.g. sugars (Hoffman 1964). In the light, photosynthesis begins and may reach a maximum in as little as 70 min in *Polytrichum commune* (Paolillo & Jagels 1969). In *Physcomitrella patens* there is no germination unless the spores are illuminated under moist conditions. Fifty per cent germination may be expected after 20 hours of light but this is not observed physically until a further 15 hours have elapsed Light is not required during the latter period. Some germination may occur after as little as 8 hours illumination while maximum percentage germination requires around 48 hours light (Cove *et al.* 1978). The promoting effect of light on moss spore germination was first discovered by Borodin, better known as a composer of music. The phytochrome system is involved since germination in *Funaria hygrometrica* is enhanced by red light and the effect is reversed by immediate exposure to far-red light (Bauer & Mohr 1959). Within a day or so of the initiation of germination in moss spores a green filament called a protonema (plural: protonemata) emerges often from the side of the spore facing the light and rhizoids grow out at the opposite end. This suggests that polarity is light-determined through mediation by phytochrome. However, in *Physcomitrella* there is no evidence that the position of protonemal emergence is orientated with respect to light. The colourless or brownish rhizoids are negatively phototropic growing downwards to attach the protonema to the substrate.

2 PROTONEMATA

2.1 STRUCTURE

(a) *Peat Mosses*

A short protonemal filament, 2–4 cells long grows from the germinating spore. The terminal cell then divides so as to give rise to a flat green plate of cells which is one cell thick. Later, marginal cells grow out so that an irregularly shaped thallus is formed while colourless filaments (rhizoids) attach it to the substratum. Eventually buds form on the margin of the protononema which develop into leafy gametophores (Anderson & Crosby 1965). Other marginal cells may grow out into additional filaments whose terminal cell can form a second thalloid protonema which will subsequently form more buds. This can happen several times so that germination of a single spore may result in quite a large clump of moss.

(b) *Granite mosses*

On germination, the protoplast of the spore divides to form a globular mass of tissue which ruptures the cell wall and then gives rise to a protonemal filament 3–5 cells long. The terminal cell then divides to produce filaments, some of which act as rhizoids, others become green, creep over the substratum expanding to form a multilobed structure, or grow upwards. Eventually buds form on the protonema (Parihar 1977).

(c) *True mosses*

The cells of the protonema which grow out from the germinating spore have many chloroplasts and cross cell walls which are perpendicular to the long axis (Fig. 6.2). Cells of this make up the primary chloronema (Plural: chloronemata). After about a week, the new cells formed by the growing protonema have fewer chloroplasts and diagonal walls between adjacent cells. Such cells make up the caulonema (Fig. 6.3). The latter develops side-branch initials, some of which form buds that give rise to leafy gametophores (Cove *et al.* 1978, 1980; Handa & Johri 1979). The protonematal filaments of mosses respond rapidly to light by curvature of terminal cells (Meyer 1948). In *Physcomitrella*

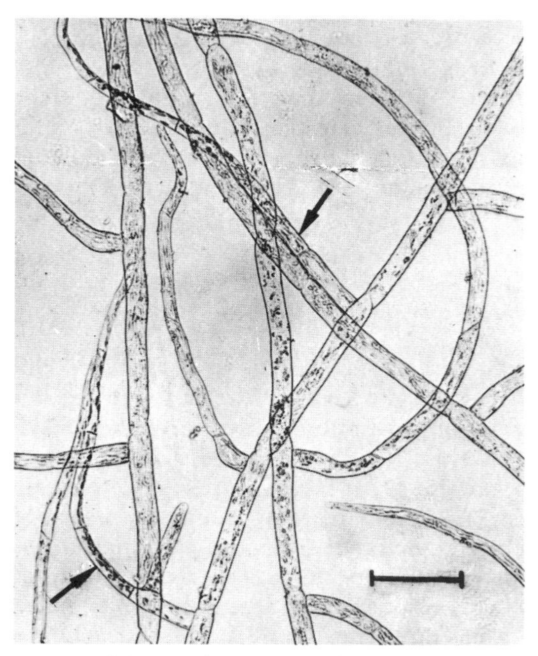

Fig. 6.3 Moss protonema of *Funaria hygrometrica* showing caulonemata with oblique cell walls and secondary chloronemata (arrows) with transverse walls and numerous chloroplasts (Handa & Johri 1979).

patens, the primary chloronema filament produced following spore germination is weakly negatively phototropic in bright light but positively phototropic in dim light (Cove *et al.* 1978). Secondary chloronemata formed from some of the branch initials on caulonemata turn towards incident light and this positive phototropic response seems to be phytochrome mediated (Cove *et al.* 1978). It is interesting that the caulonemata grow at right angles to light.

The protonemal stage in some mosses becomes quite extensive before gametophore buds develop. It is usually filamentous but is thalloid in quite a number of true mosses including *Tetraphis* and *Diphyscium* (Engler & Prantl 1924). The filamentous protonemal stage of *Viridivellus pulchellum* from Australia is so dominant and conspicuous that a growth resembling a tiny green fleece is formed on the soil surface. The gametophore is reduced to about four tiny perichaetial leaves whose function is to protect the sex organs and capsules which arise from fertilized archegonia (Stone 1976). An extensive protonema is also found in the most remarkable moss *Schistostega pennata* which grows on the floors of small caves or animal burrows. On looking into a cave colonized by this moss, innumerable golden-green points of light are seen to sparkle and gleam. This is caused by reflection of incident light from the lens-like cells of the extensive protonema. This forms a weft over the surface of the ground from which upright growing filaments develop bearing groups of spherical cells like bunches of grapes. Their shape is such that light is focused onto a spot at the rear of each cell where there is a group of chloroplasts. These absorb only part of the light, the remainder being reflected back as from a concave mirror giving the protonema its luminous appearance. The luminous glow is sometimes called goblin gold (Kerner & Oliver 1897; Crum 1973). Similar protonemata are found in *Mittenia plumula* from Australia which occupies

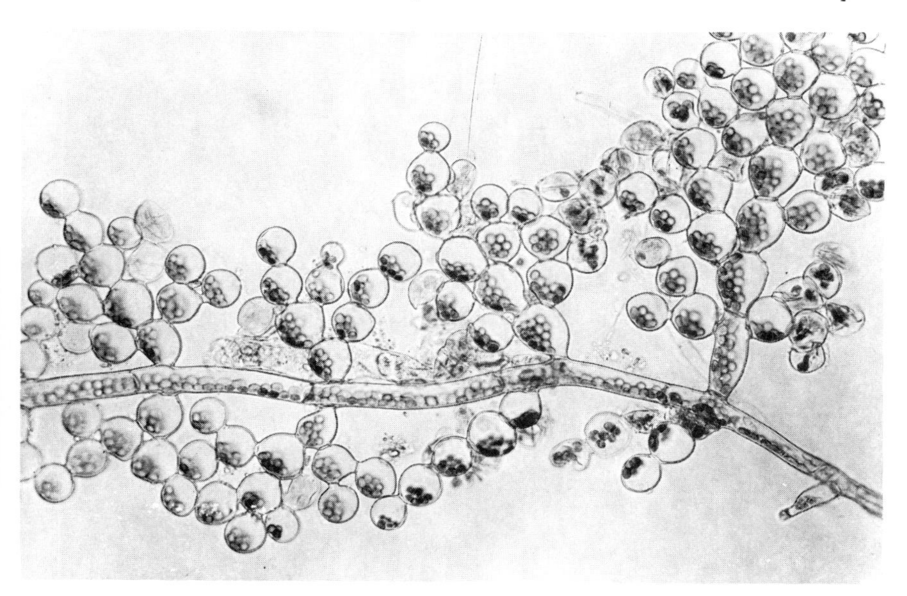

Fig. 6.4 Groups of spherical cells on the protonema of *Mittenia plumula* which reflect part of the incident light. This moss is from Australia but a similar feature is shown by *Schistostega pennata* from the northern hemisphere (Stone 1961).

the same ecological niche (Stone 1961; Fig. 6.4). In these mosses, although small numbers of buds develop which form leafy gametophores, the main assimilatory tissue is the protonema.

The morphogenetic change from chloronemata to caulonemata to leafy gametophores in the gametophyte generation of mosses has intrigued bryologists. It has been the subject of research for a long time, much of the early work being done in Germany (see Bopp 1968). Recent studies are summarized in the following sections.

2.2 DIFFERENTIATION

(a) *Primary chloronema to caulonema*

A series of mutant strains of *Physcomitrella patens* have been produced using mutagenic chemicals. Such mutants are often defective at synthesizing their own plant hormones. By studying these mutants it has been shown that the change from chloronema to caulonema requires low levels of the plant hormone auxin. In addition normal strains of the moss will not produce caulonema if grown under a constant stream of fresh medium. This is due to leaching out of the auxin (Ashton *et al.* 1979 a, b; Cove *et al.* 1980). As might be suspected, the proportion of caulonema to chloronema may be increased by supplying the protonema with additional amounts of auxin and this has been shown in *Funaria hygrometrica* (Johri & Desai 1973) and *Physcomitrella patens* (Cove *et al.* 1980).

The changes which accompany differentiation from chloronema to caulonema are both structural (cross cell wall orientation and chloroplast frequency) and biochemical. Thus specific caulonemal proteins are synthesized in several mosses (Sood *et al.* 1978) and at least in *Funaria hygrometrica*, the caulonemal cells which are not destined to divide further or form buds exhibit increased DNA contents (8 to 16 times normal) (Knoop 1978). This correlates with the observation that the nuclei of many caulonemal cells are larger than those found in the corresponding chloronema. The value or function of this increase in DNA content is unknown. However, if such cells are induced to become active and divide by isolating them from the rest of the protonema, they maintain their high DNA content.

(b) *Caulonema to secondary chloronema*

The cells comprising the caulonemal filaments divide to form side branch initials, many of which develop into secondary chloronemata while a few form buds giving rise to gametophores. The secondary chloronemata morphologically resemble the primary chloronemata in having many chloroplasts and cross cell walls that are perpendicular to the long axis. In *Physcomitrella patens* the formation of secondary

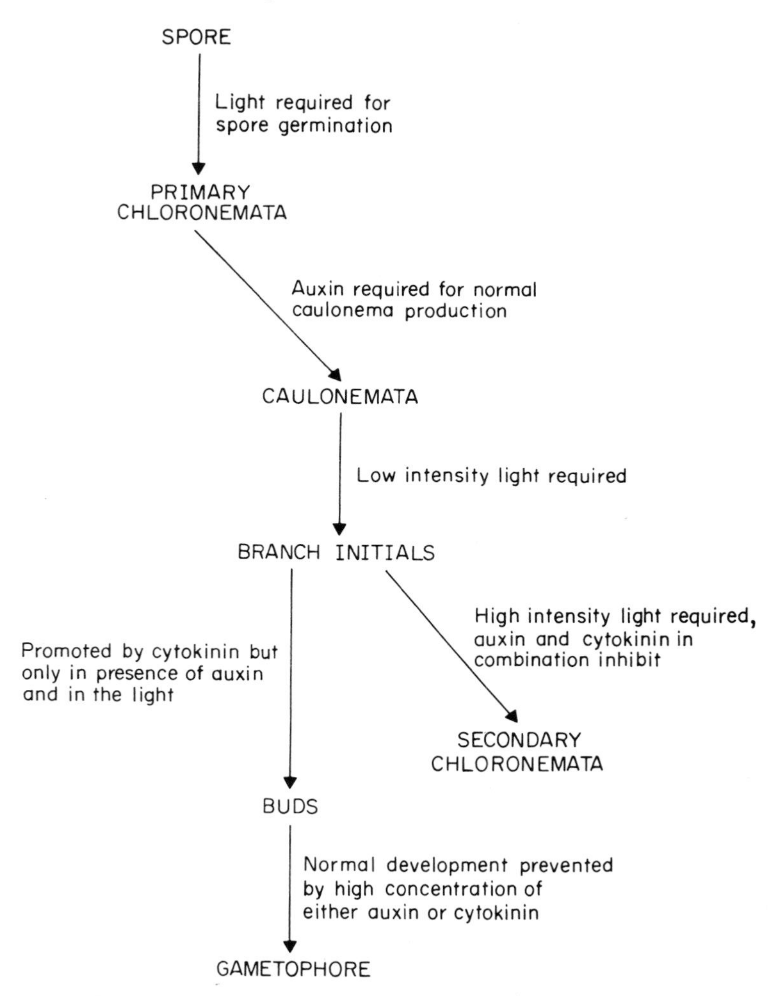

Fig. 6.5 A model that accounts for the effects of auxin and cytokinin during the differentiation of gametophores from protonema in the moss *Physcomitrella patens* (adapted from Cove *et al.* 1980).

chloronemata is enhanced by high light intensities but inhibited by auxin and cytokinin (Fig. 6.5) (Cove *et al.* 1980). In *Funaria hygrometrica*, formation of secondary chloronemata is promoted by cyclic adenosine monophosphate (cAMP). This nucleotide is also involved in differentiation and morphogenesis in fungi and slime molds. Induction of secondary chloronemata in *Funaria hygrometrica* is especially favoured when the ratio of cAMP to auxin is greater than 20. High cAMP

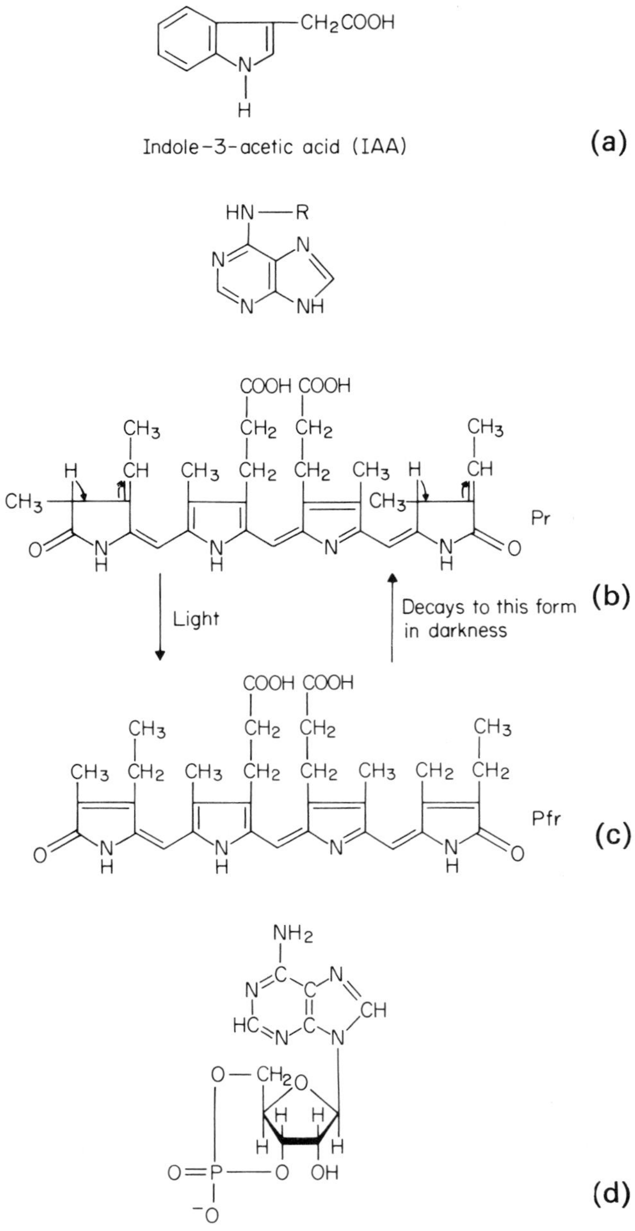

Indole-3-acetic acid (IAA)

(a)

Pr

Light

Decays to this form in darkness

(b)

Pfr

(c)

(d)

Fig. 6.6 The plant hormones and growth factors involved in the differentiation of moss protonemata (Wareing & Phillips 1978).
(see Legend opposite).

levels therefore maintain the chloronemal phase while an increase in auxin and/or fall in cAMP levels results in development of caulonemata. cAMP levels in chloronemata have been found to be 4–7 times higher than those in caulonemata (Handa & Johri 1979).

The important role played by auxin in protonemal differentiation has been demonstrated using mutants of *Physcomitrella patens* which are defective in the production of this plant hormone. Such mutants only form chloronemata unless supplied exogenously with auxin. They are then able to develop caulonemata (Ashton *et al.* 1979a).

(c) *Caulonema to gametophore*

Two plant hormones, auxin and cytokinin (Fig. 6.6), are needed for the formation of gametophore buds on the caulonema of *Physcomitrella patens*. Cytokinin causes a massive increase in the production of buds but the ability to respond to cytokinin is dependent on there being adequate levels of auxin. It is suggested that cytokinin both depresses the growth of secondary chloronemata and induces bud formation in the branch initials on the caulonemata. This has again been demonstrated using mutants which are unable to synthesize proper amounts of auxin or cytokinin. Light is also required for the synthesis of an additional essential component which decays rapidly in the dark. Thus cytokinin will not induce bud formation on cultures grown in the dark or those transferred to darkness for several days before treatment (Ashton *et al.* 1979a, b). In *Funaria hygrometrica*, it appears that cytokinin and light alone are sufficient for bud induction while only cytokinin seems necessary in *Ceratodon purpureus* as protonemata of this moss will form buds in the dark (Spychala *et al.* 1975). Cyclic AMP is ineffective in gametophore bud induction and not involved in this stage of protonemal differentiation (Schneider *et al.* 1975).

The efficacy of cytokinin in bud induction is exemplified by a recent report that buds can be induced to form almost immediately after spore germination in *Polytrichum juniperinum* by incubating spores in the light on a medium containing 10^{-6} M cytokinin (as benzylaminopurine) and 1% glucose (Nehlsen 1979). The target organs for cytokinin action

---

(a) Auxin.
(b) Cytokinins.
(c) Phytochrome (This protein has a molecular weight of 120 000 daltons.
The light absorbing non-protein part is thought to undergo the indicated change in response to light).
(d) Cyclic adenosine monophosphate.

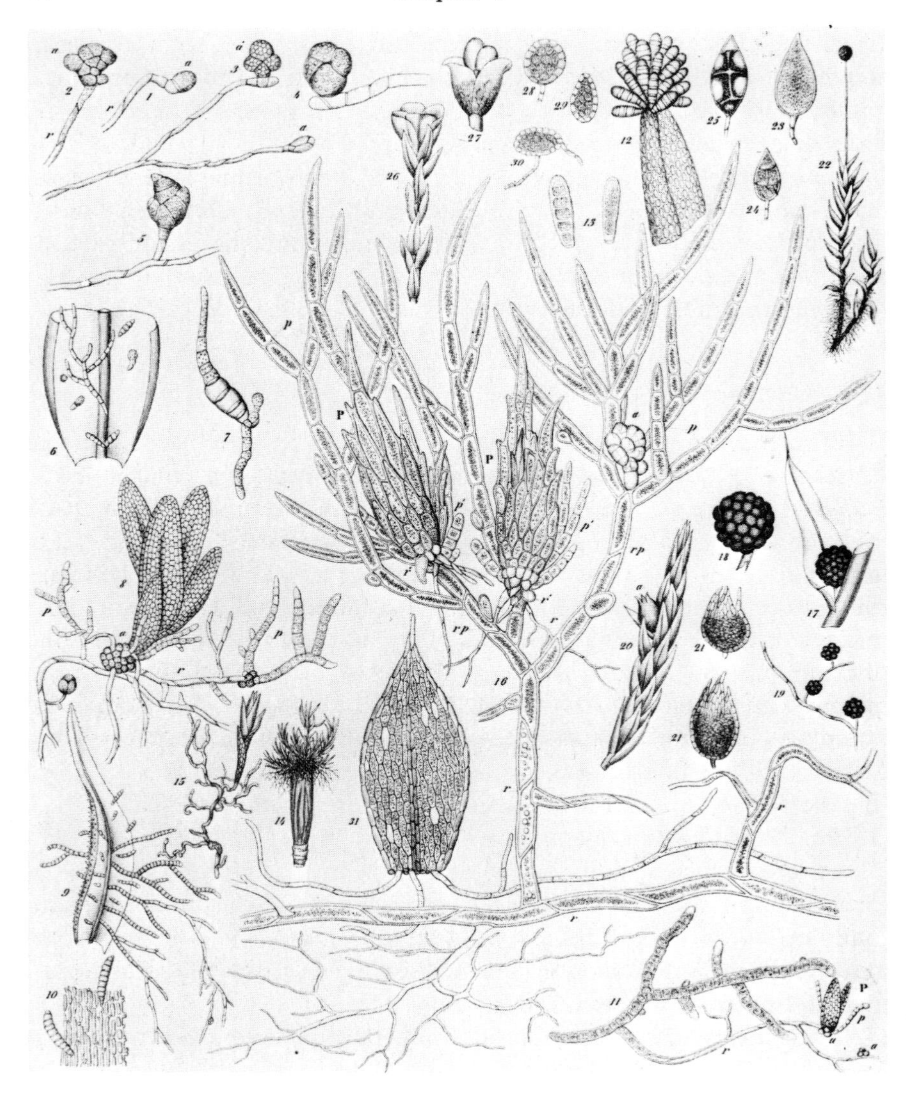

Fig. 6.7 The diverse means of vegetative propagation found in mosses (Schimper 1860). (1–5) buds on the protonema of *Funaria hygrometrica*, (6) protonema developing from a leaf of *Orthotrichum obtusifolium*, (7–8) young plants developing on adventitious protonema from the leaf, (9) Filaments growing from a leaf of *Orthotrichum lyellii* some developing into typical protonemata, (10) part of same leaf, (11) a filament developing into a protonema which in turn has buds and a young plant, (12–13) gemmae on the leaf tip of *Ulota phyllantha*, (14) young plants arising on tufts of pale protonema on the apices of leaves at the top of *Leucobryum glaucum* plants, (15) a single plant together with the adventitious protonema from which it has developed, (16) Young plants (p) of *Ephemerum serratum* developing from an underground

are the side branch initials on the caulonemata. A decrease in the activity of an enzyme, ribonuclease, is observed about four hours after administration of cytokinin to protonemata of *Ceratodon purpureus* or *Funaria hygrometrica*. This decrease lasts for at least 10 hours and is thought to be due to inhibition of the synthesis of the enzyme (Spychala *et al.* 1976). A similar fall in ribonuclease activity is also found during the spontaneous formation of gametophore buds on the protonema of these mosses (Spychala *et al.* 1975). Presumably, during the period immediately after differentiation is triggered by natural or exogenous cytokinin, important messenger ribonucleic acid and transfer ribonucleic acid is synthesized. Inhibition of ribonuclease formation during this time ensures that the essential functions of this ribonucleic acid will not be interfered with due to degradation by the enzyme. Later as the buds develop, greatly increased activity of various enzymes including ribonucleases and peroxidases is observed, the cells being metabolically very active (Spychala *et al.* 1976; Sobkowiak *et al.* 1976). Finally as the leafy gametophores are formed, they become positively phototropic and this response is mediated by phytochrome (Cove *et al.* 1978).

In conclusion, the differentiation of leafy gametophores from the first formed protonemata evidently requires the endogenous synthesis of several hormones and factors (Fig. 6.6). These play important and interdependent roles in the several steps of development (Ashton *et al.* 1979a).

(d) *Protonemata developing on gametophore tissue*

Protonematal filaments may develop from almost any part of a moss plant, providing conditions are suitable, and give rise subsequently to new plants. Such protonemata are sometimes called secondary protonemata and their formation is promoted by a moist environment (Fig. 6.7). If a moss tuft is inverted and kept damp then protonemata may develop from the rhizoids and hundreds of new gametophore buds can develop.

---

protonema (r). Transformation of the protonema (caulonema) into prothallial filaments (chloronema) is conspicuous at (rp) and the latter bear buds (a), (17–18) Scarlet gemmae in the axils of the leaves in *Bryum rubens*, (19) rhizoids of the same moss bearing red gemmae. (20) bulbils in leaf axils of *Pohlia annotina*, (21) a single bulbil, (22) a plant of *Aulacomnium androgynum* with a 'drum stick' shaped gemmae head, (23–25) individual gemmae, (26–27) a sterile stem of *Tetraphis pellucida* bearing a gemma cup, (28–29) individual gemmae, (30) a germinating gemma, (31) a deciduous leaf of *Funaria hygrometrica* with secondary protonema emanating from the basal cells.

If the tuft is inverted once more, protonemata often grow out from the original stems which may be moribund and new buds again form. Continual repetition of this process results in spherical colonies termed moss balls. Species which have the ability to form moss balls are able to colonize unstable substrates and may thus have an advantage over competing plants. This growth form is not limited to mosses but is known for other plants including lichens (Weber 1977). Moss balls have been found, usually in temperate and subarctic climates, in forests,

Fig. 6.8 Polsters of *Drepanocladus berggrenii* on the ablating ice of Gilkey Glacier, Alaska. The polsters give the glacier the appearance of having a cobble strewn surface (Heusser 1972).

lakes, sand dunes, and gently sloping ice fields (Fig. 6.8). In the latter instance, the moss balls which are termed polsters or glacier mice may form the start of a succession which ends in a forest overlying many feet of slowly melting ice (Heusser 1972). Wind and water are primary factors causing the moss balls to move although in forests the scratching of game birds (pheasants in England, wood hens in New Zealand) may also be important (Shacklette 1966). Some 15 species of mosses including *Andreaea*, *Drepanocladus*, *Leucobryum* and *Rhacomitrium* are reported to form moss balls which invariably consist of a single species.

Capsules have not been seen so that sexual reproduction appears to have been replaced by colony fragmentation in ecotypes that form moss balls.

Many mosses can reproduce vegetatively by fragments of the parent colony breaking off and being carried by wind or water to suitable locations where the fragments generate secondary protonemata which bear new buds. Under ideal conditions, even detached leaves can form new plants via the production of protonemata. The various parts of a moss gametophore are under control of hormones, especially indole acetic acid (auxin), produced by the apical cell at the growing tips of the plant. The hormones inhibit the growth of buds in the axils of leaves lower down the stem and prevent regeneration. Once released from this control, regeneration begins from detached parts and is mediated by the phytochrome system; red light inducing the process and far red light inhibiting it. Within 48 hours of detachment, several cells on the surface of a moss leaf, e.g. in *Mnium affine*, will contain dividing chloroplasts. The cells become meristematic and grow out into secondary protonemata which can subsequently form buds and new gametophores (Giles 1971). Over 60% of leaves, detached from the apical green region of *Pleurozium schreberi* shoots, developed secondary protonemata when placed on a sand-peat mixture or on moist filter paper. Buds were formed subsequently on many of the protonemata. This regenerative ability accounts for the common occurrence of this moss in Britain where the rarity of male plants results in very few capsules being produced (Longton & Greene 1979). However some mosses possess a more specialized method of vegetative reproduction in the form of gemmae. These are multicellular plates of tissue on a short stalk which form at various locations on the parent gametophore depending on the species. Gemmae, like protonemata can also form on the rhizoids of mosses and are often subterranean. The gemmae stalks usually break easily and the propagules are dispersed by wind or rain. Rhizoids and protonemata develop on gemmae which lodge in a suitable place and buds form on the latter so forming a new moss. Gemmae are most common on species of moss which grow on trees or colonize soil. In the former case, the gemmae can be splashed from branch to branch, probably also being transported passively on insects and birds (Scharf 1978). In the soil mosses, the subterranean gemmae are termed tubers (Whitehouse 1973, 1978) and may remain dormant until exposed to light, or provide a means for the moss to survive dry or cold periods (Sachs & Vines 1882; Watson 1964). Indeed *Tortula amplexa* is thought to have been introduced to Britain from western North America by means of rhizoid gemmae (tubers). The moss was discovered in Britain growing in a polythene bag on moist clay which came from a quarry in Leicestershire, England, and a visit revealed that the moss grew on disturbed sites there. As only female plants were found it is presumed that the moss has spread from a single propagule (tuber) introduced perhaps in soil on the shoe of an American visitor (Side & Whitehouse 1974).

Mosses have been said to exhibit more copious and varied means of vegetative reproduction than any other section of the plant kingdom (Sachs & Vines 1882). These usually involve the growth of secondary protonemata at some stage (Fig. 6.7). Evolution of the vegetative methods of reproduction presumably helps to compensate for the problems (mentioned in Chapter 4) associated with fertilization by a swimming sperm cell in a terrestrial environment.

(e) *Protonemata developing from sporophyte tissue*

If moss capsules or particularly segments of capsule stalks are kept moist, they can be induced to produce protonemata fairly easily in the laboratory. This was first discovered by Pringsheim in 1876 who, to his surprise, found that such protonemata formed gametophores. Subsequently it was shown that the gametophores were diploid. This formation of gametophores direct from sporophyte tissue without spore formation (during which meiosis occurs) is called apospory (Smith 1978).

The above mechanism provides one way in which polyploid populations of mosses (i.e., those with more than one set of chromosomes) may have evolved (see Chapter 4). There is a report of regeneration of gametophores from a capsule of *Funaria* which had been buried accidentally (Crum 1973) but this seems to be the only instance where the production of protonemata and gametophores has been observed from sporophyte tissue under natural conditions (Smith 1978).

3 REFERENCES

ANDERSON L.E. & CROSBY M.R. (1965) The protonema of *Sphagnum meridense* (Hampe) C. Muell. *Bryologist*, **68**, 47–54.

ASHTON N.W., GRIMSLEY N.H. & COVE D.J. (1979a) Analysis of gametophytic development in the moss *Physcomitrella patens* using auxin and cytokinin resistant mutants. *Planta (Berl.)*, **144**, 427–35.

ASHTON N.W., COVE D.J. & FEATHERSTONE D.R. (1979b) The isolation and physiological analysis of mutants of the moss, *Physcomitrella patens*, which over-produce gametophores. *Planta (Berl.)*, **144**, 437–42.

BAUER L. & MOHR H. (1959) Der Nachweis des reversiblen hellrot-dunkelrot-reactions-systems bei Laubmoosen. *Planta (Berl.)*, **54,** 68–73.

BOPP M. (1968) Control of differentiation in fern allies and bryophytes. *Ann. Rev. Pl. Physiol.* **19,** 361–77.

BOROS A. & JARAI-KOMLODI M. (1975) An atlas of recent European moss spores. Akaemiai Kiado, Budapest.

COVE D.J., SCHILD A., ASHTON N.W. & HARTMAN E. (1978) Genetic and physiological studies of the effect of light on the development of the moss *Physcomitrella patens. Photochem. Photobiol.* **27,** 249–54.

COVE D.J., ASHTON N.W., FEATHERSTONE D.R. & WHANG T.L. (1980) The use of mutant strains on the study of hormone action and metabolism in the moss *Physcomitrella patens*. The plant genome. *Proc.* 4th *Symp. John Innes Instit.* Norwich 1979.

CRUM H. (1973) Mosses of the Great Lakes Forest. *Contrib. Univ. Mich. Herb.* **10,** 1–404.

ENGLER A. & PRANTL K. (1924) Die Naturlichen Pflansenfamilien nebst Ihren gattungen und wichtigeren arten insbesondere den nutzpflanzen. *Verlag, Leipzig,* **10,** 7–21.

GILES K.L. (1971) Differentiation and regeneration in Bryophytes: A selective review. *N.Z. J. Bot.* **9,** 689–94.

HANDA A.K. & JOHRI M.M. (1979) Involvement of cyclic adeonosine-3′, 5′-monophosphate in chloronema differentiation in protonema cultures of *Funaria hygrometrica. Planta (Berl.),* **144,** 317–24.

HEITZ E. (1942) Die kiemende *Funaria*—spore als physiologisches versuchobjekt *Ber. Dtsch Bot. Ges.* **60,** 17–27.

HEUSSER C.J. (1972) Polsters of the moss *Drepanocladus berggrenii* on Gilkey Glacier, Alaska. *Bull. Torrey Bot. Club.* **99,** 34–6.

HOFFMAN G.R. (1964) The effects of certain sugars on germination in *Funaria hygrometrica* Hedw. *Bryologist,* **67,** 321–9.

JOHRI M.M. & DESAI S. (1973) Auxin regulation of caulonema formation in moss protonema. *Nature New Biol.* **245,** 223–4.

KERNER A. & OLIVER F.W. (1897) *The Natural History of Plants.* Blackie, London.

KNOOP B. (1978) Multiple DNA contents in the haploid protonema of the moss *Funaria hygrometrica* Sibth. *Protoplasma,* **94,** 307–14.

LONGTON R.E. & GREENE S.W. (1979) Experimental studies of growth and reproduction in the moss *Pleurozium schreberi* (Brid.) Mitt. *J. Bryol.* **10,** 321–38.

MEYER S.L. (1948) Physiological studies on mosses. 7. Observations on the influence of light on spore germination and protonemal development in *Physcomitrium turbinatum* and *Funaria hygrometrica. Bryologist,* **51,** 213–17.

MUELLER D.M.J. (1974) Spore wall formation and chloroplast development during sporogenesis in the moss *Fissidens limbatus. Am. J. Bot.* **61,** 525–34.

NEHLSEN W. (1979) A new method for examining induction of moss buds by cytokinin. *Am. J. Bot.* **65,** 601–3.

OLESEN P. & MOGENSEN G.S. (1978) Ultrastructure, histochemistry and notes on germination stages of spores in selected mosses. *Bryologist,* **81,** 493–516.

PAOLILLO D.J. & JAGELS R.H. (1969) Photosynthesis and respiration in germinating spores of *Polytrichum. Bryologist,* **72,** 444–51.

PARIHAR N.S. (1977) Bryophyta. In *An Introduction to Embryophyta,* Vol. 1, 5th Ed. Central Book Depot. Allahabad.

PURI P. (1973) The Bryophyta: a Broad Perspective. Atma Ram. Dehli.

SACHS J. & VINES S.H. (1882) Text-book of botany. 2nd ed, pp. 367–70. Clarendon Press, Oxford.

SCHARF C.S. (1978) Birds and mammals as passive transporters for algae found in lichens. *Can. Field-Nat.* **92,** 70–1.

SCHIMPER W.P. (1860) Icones morphologicae atque organographicae introductionem synopsis muscorum europaeorum. E. Schweizerbart, Stuttgartiae.

SCHNEIDER J., SZWEYKOWSKA A. & SPYCHALA M. (1975) Evidence against mediation of adenosine-3′, 5′- adenosine-3′, 5′-cyclic monophosphate in the bud inducing effect of cytokinins in moss protonema. *Acta Soc. Bot. Pol.* **44,** 607–13.

SHACKLETTE H.T. (1966) Unattached moss polsters on Amchitka Island, Alaska. *Bryologist,* **69,** 346–52.

SIDE A.G. & WHITEHOUSE H.L.K. (1974) *Tortula amplexa* (Lesq.) Steere in Britain. *J. Bryol.* **8,** 15–18.

SMITH A.J.E. (1978) Cytogenetics, biosystematics and evolution in the Bryophyta. In *Advances in Botanical Research,* ed. H. Woolhouse, Vol. 6, pp. 196–277. Academic Press, London.

SOBKOWIAK A., MLODZIANOWSKI F. & SZWEYKOWSKA A. (1976) Cytochemical localization of peroxidase activity in the protonema of the moss *Ceratodon purpureus* during differentiation of gametophore buds by kinetin. *Acta Bot. Soc. Pol.* **45,** 119–25.

SOOD S., BRENNER K. & BOPP M. (1978) The occurrence of caulonema-specific proteins in different moss species. *Planta (Berl.),* **138,** 299–301.

SPYCHALA M., SCHNEIDER J. & SZWEYKOWSKA A. (1975) Relationship between formation of gametophore buds in the protonema of mosses and increase in ribonuclease activity. *Acta Soc. Bot. Pol.* **44,** 433–41.

SPYCHALA M., KORCZ-ZAJCHERT I. & SZWEYKOWSKA A. (1976) Cytokinin-induced decrease in ribonuclease activity and initiation of gametophore buds in the protonema of mosses. *Acta Soc. Bot. Pol.* **45,** 327–33.

STONE I.G. (1961) The highly refractive protonema of *Mittenia plumula* (Mitt.) Lindb. (Mitteniaceae). *Proc. Roy. Soc. Victoria,* **74,** 119–25.

STONE I.G. (1976) A remarkable new moss from Queensland Australia—*Viridivellus pulchellum* new genus and species (new family Viridivelleraceae) *J. Bryol.* **9,** 21–31.

WAREING P.F. & PHILLIPS I.D.J. (1978) The control of growth and differentiation in plants. 2nd Edition. Pergamon Press, Oxford.

WATSON E.V. (1964) *The Structure and Life of Bryophytes,* pp. 98–103. Hutchinson, London.

WEBER W.A. (1977) Environmental modification and lichen taxonomy. In *Lichen Ecology,* ed. Seaward M.R.D., pp. 8–29. Academic Press, London.

WHITEHOUSE H.L.K. (1973) The occurrence of tubers in *Pohlia pulchella* (Hedw.) Lindb. and *Pohlia lutescens* (Limpr.) Lindb. fil. *J. Bryol.* **7,** 533–40.

WHITEHOUSE H.L.K. (1978) *Bryum cruegeri* Hampe: a tuber bearing moss new to Africa. *J. Bryol.* **10,** 113–15.

CHAPTER 7
MOSS–ANIMAL ASSOCIATIONS

The biomass of mosses in many vegetational zones of the world is considerable. However it is in the tundra that they are most abundant. On over three million square kilometres of Russian tundra approximately 50% of the above ground plant biomass consists of moss while in the wet sedge meadows of the Western Taimyr, the figure is in excess of 90%. In spite of this high biomass, annual production of mosses is about 10% of that in the higher plants growing in the same area and varies from 1–50 g dry wt/m^2/yr. However in favourable areas production may be 50–95% that of the higher plants and may reach 1000 g dry wt/m^2/yr (Clarke *et al.* 1971; Kallio & Karenlampi 1975; Oechel & Sveinbjornsson 1978). This large amount of material should provide the potential for diverse moss–animal associations and yet few have been described and quantitative data are scarce.

1 NUTRITIVE VALUE

1.1 CHEMICAL COMPOSITION

The caloric content of mosses from the Canadian tundra has been estimated to vary from about 4·5 to 5·0 kcal/g (when corrected for ash) with aquatic species exhibiting the higher value. Mosses therefore have comparable caloric contents to higher plants (arctic willow, herbs and sedges) growing in the same habitat (Pakarinen & Vitt 1974). Flowering plants are comparatively rich in protein and fats (*c.* 15% and 5% respectively). In contrast, mosses have large amounts of lignin-like compounds (comparable to woody stems) but a total protein content of only about 4% and total fats of only 2%. The amount of sugars in mosses varies from about 1–5% depending on the species and they are reported to be quite high in vitamins, especially B_2 (Margaris & Kalaitzakis 1974; Sugawa 1960).

1.2 Palatability

While one cannot assume that palatability to man has any relevance in considerations of other animals, it is worth noting that fresh moss is, to us, unpleasant. The taste of most species is rather like raw green beans and some such as *Dicranum* have a strong rather peppery nature. *Rhodobryum giganteum* is sickeningly sweet when fresh and requires rinsing from the mouth (Crum 1973).

Dried moss seems to be much more palatable and it has been found that powdered *Barbella pendula* can be added to the food of chickens and puppies with no distaste or ill effects; indeed the animals thrive (Sugawa 1960).

2 VERTEBRATE CONSUMERS

Mosses have been identified in the alimentary tract of bison (Ukraintseva *et al.* 1978). The situation may be comparable to reindeer which do not appear to digest consumed moss easily although their enzymes readily break down lichens. The high content of lignin-like substances in mosses may account for this. Recently it has been found that while reindeer in continental ranges of Europe and North America feed principally on lichens during winter, animals in the high arctic consume mosses and higher plants. For example reindeer of the Svalbard Archipelago, Spitzbergen, feed heavily on mosses and exhibit physiological adaptations to their extreme environment accumulating large quantities of fat and conserving energy through reduced activity, especially in summer. By these means, they have a greater population density and faster growth rate than animals further south in spite of there being less plant biomass available (Reimers & Klein 1979).

At Point Barrow, Alaska, up to 40% of the winter diet of lemmings may be composed of mosses (Collins & Oechel 1974; Oechel & Sveinbjornsson 1978) while the frequency of moss in the diet items found in the stomach of bank voles from the forests of northern Sweden was close to 20% (Hansson 1979). There is also the possibility that dispersal of *Voitia hyperborea*, a member of the Splachnaceae, depends on the moss capsules being eaten by musk oxen and caribou. This may happen accidentally as they graze other plants. The moss grows on the dung of the animals but the capsules do not open so that dissemination of the spores to fresh dung appears to require the capsules being eaten by mammals or insects (Steere 1974).

Appreciable amounts of moss are apparently eaten by barnacle geese, before other food plants become available, when they arrive in Spitzbergen, Scandinavia, at the end of the spring migration. Mosses are eaten intensively during the prelaying and laying periods but also in the summer to a lesser extent. Thus 27 of 28 samples of adult and gosling droppings at the latter time contained moss and the mean area of moss leaves as a percentage of all plant epidermis was close to 25% (Owen pers. comm.). Pink-footed geese in Iceland regularly consume mosses but they do not form a high proportion of the overall diet (Gardarsson *et al.* 1972). Snow geese also take considerable quantities of moss in spring and early summer, possibly accidentally, as they probe for the roots on which they principally feed.

The capsules of mosses are also eaten by a few birds; Norwegian grouse chicks are reported to feed on capsules of *Bryum* and *Polytrichum*. The capsules of moor-inhabiting mosses may also be found in over half of the crops of red grouse chicks from Scotland. Other than young shoots and flowers of heather, moss capsules apparently constitute the main food for the chicks (Lid & Meidell 1933). In addition, *Dicranoweisia cirrata* capsules provide a food item for blue tits and marsh tits in woodland areas in Britain (Betts 1955).

3 INVERTEBRATE CONSUMERS

A wide range of invertebrates use mosses as a food source, and many have become specialized to live in mosses, even in *Grimmia pulvinata* which forms tufts on house roofs. The animals inhabiting this moss include tardigrades, rotifers, nematodes and amoebae (Corbet & Lan 1974). Moss cushions on trees in damp woodlands or on moist ground may contain scorpion flies (Fig. 7.1) whose larvae, in the case of *Boreus* sp. and *Caurinus* sp., also consume moss tissue (Penny 1977; Russell 1979; Shorthouse 1979). In other situations mosses form the food source for aphids, fly larvae, caterpillars of small moths, small grasshoppers and some gastropods (Gerson 1969; Crum 1973; Dickinson & Maggs 1974). Details of some of these relationships are provided below.

Fig. 7.1 An adult snow scorpionfly, *Boreus brumalis*. This small insect which is up to 5 mm long lives and feeds on mosses. In Northern Ontario the adults emerge on the snow in spring to mate. Eggs are laid at the base of mosses on which the larvae feed. (Shorthouse 1979).

3.1 Grasshoppers

The value of mosses to these animals is derived from the following evidence. Some 80% of the faecal pellets of *Tetrix undulata* and *Tetrix subulata* (ground hoppers) is made up from the leaves of *Hypnum* sp. and related mosses. The remainder consists of stems, protonemata and rhizoids of mosses together with epidermal tissue of higher plant leaves (Verdcourt 1947). Mosses have also been found in the crops of *Melanoplus femur-rubrum* and *Dissosteira carolina*. The latter species, as well as *Melanoplus sanguipes* and *Schistocerca lineata*, have been observed eating mosses, possibly for the water they contain (Uvarov 1977).

In food trials on a number of Bavarian grasshoppers, all except *Tetrix* rejected mosses. A member of this genus, probably *Tetrix tenuicorne*, consumed more than half of the offered sample of the following mosses: *Barbula convoluta*, *Bryum argenteum*, *Ceratodon purpureus* and *Funaria hygrometrica* while other species were eaten in smaller amounts, The tiny grasshoppers in this genus are remarkable in being able to swim under water. They consume relatively large amounts of food, eating frequently, and die if deprived of food for more than a few days (Kaufmann 1965; Ragge 1965).

3.2 Moths

The caterpillars of some of the species in the primitive microlepidopteran family, Micropterygidae, subsist on mosses. These small moths generally have buff and purple wings, a dull golden body, and a wingspan of about one centimetre. The caterpillars of one, *Micropteryx calthella* are reported to eat *Eurhynchium praelongum*, *E. swartzii* and *Thuidium tamariscinum* and grow to about 3 mm long (Chapman 1894).

A number of different moss caterpillars were found in moss cushions in North America. Some were of irregular shape, patterned green and brown, and were thought to belong to the Noctuidae. Others were reddish-brown up to 8 mm long, sparsely haired with glossy brown heads and were found during a study of craneflies. These caterpillars are thought to have given rise to the light grayish moths (found in the cranefly rearing dishes) which have been identified as *Crambus alboclevellus*, *C. elegans* and *C. albellus* (Byers 1961).

3.3 Flies

One theory on the origin of flies as a group, suggests that they arose from ancestors which, though terrestrial, had larvae that lived in wet moss (Oldroyd 1964). Some of the craneflies (the Tipulidae) are amongst the most ancestral of the living fly families.

(a) *Aquatic mosses*

The larvae of the primitive cranefly genera *Triogma* and *Phalacrocera* live among moss submerged in moving water. For example, *Triogma trisulcata* lives in mountain streams where its larvae are attached to stems of *Fontinalis antipyretica*. The larvae have leaf-like appendages on their bodies so that they resemble the moss on whcih they live (Alexander 1920). Other species have finger-like processes which become covered with algae and other organisms. These presumably make the larvae less attractive or less easily seen by predators. Other larvae which live on aquatic mosses include the mayflies *Baetis rhodani* which consume moss, detritus and algae and *Ephemerella notata* which subsists chiefly on *Fontinalis* and the filamentous alga *Ulothrix*. The Dipteran larva of *Simulium tuberosum* is another which feeds on aquatic moss though it also consumers leaves or filamentous algae (Jones 1949). Other Diptera

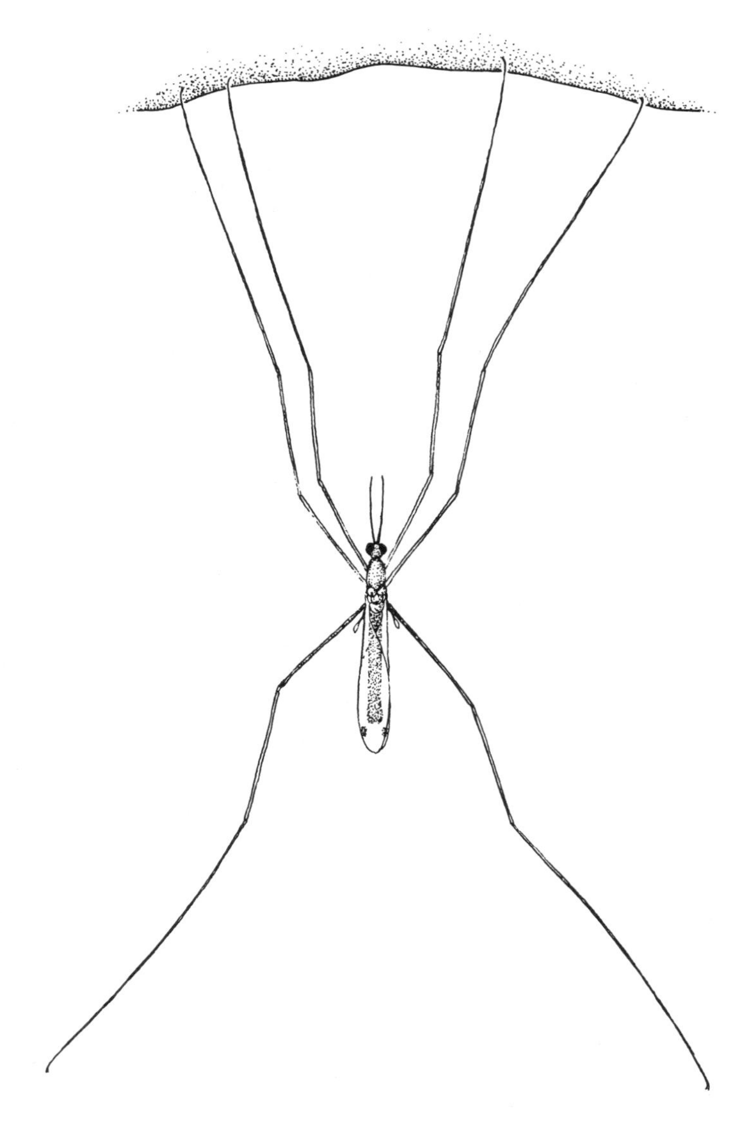

Fig. 7.2 An adult crane fly of the genus *Oropeza* whose larvae live on mosses (Byers 1961).

in the subfamily Hemerodromiinae have larvae which live on aquatic or subaquatic mosses (Brindle 1964).

(b) *Terrestrial mosses*

The larvae of craneflies belonging to the genus *Dolichopeza* (subgenera *Dolichopeza* and *Oropeza*) are moss dwellers and an excellent study has been made on them in the U.S.A. (Byers 1961). The adult insects which are commonly found in cool, wooded areas, have very long legs and a body about 15 mm long with a wingspan about twice this (Fig. 7.2). They hang by their legs in shaded situations and periodically take off on a peculiar dancing flight. After mating, the female flies close to clumps or carpets of moss and commences a series of bouncing motions up to 90 times a minute during which time she deposits her 100 or so eggs.

The larvae (Fig. 7.3) vary in colour, for example, *Dolichopeza americana* is green with dark lines and blotches which break up the outline of the individual. This correlates with its feeding habit in that this species browses at the surface of the moss cushion in bright light. *Oropeza* larvae are uniformly brown or green and, as might be conjectured, conceal themselves in the moss cushion during the day, emerging to feed at night. Larvae belonging to *Dolichopeza* and *Oropeza* have been found in a wide variety of mosses (*Sphagnum*, *Tetraphis*, *Orthotrichum* and different genera in the Hypnaceae). Aquatic mosses and those from very dry habitats seem to be avoided by these insects. Often larvae of several different species of these craneflies may be found in a single moss clump less than 10 cm in diameter but almost never seem to cause significant damage. They continue feeding until they are up to 16 mm long, moulting three times: finally they pupate and about a week later emerge as adults. In cold regions, one generation per year is normal but in more temperate parts of America (New England and the south) two generations occur. Bryophytes also harbour craneflies of other genera including *Tipula*, *Liogma*, *Limonia*, *Erioptera* and *Gonomyia*, some of whose species derive their food from these plants (Byers 1961).

Another group of interesting flies are the gall midges. These generally form galls which may be harmful or unsightly to higher plants. Some of the larvae of this group, the Cecidomiidae, do not form galls but follow the ancestral habit of feeding on detritus mosses and fungi. Any larvae of a featureless pink colour in a moss cushion may be

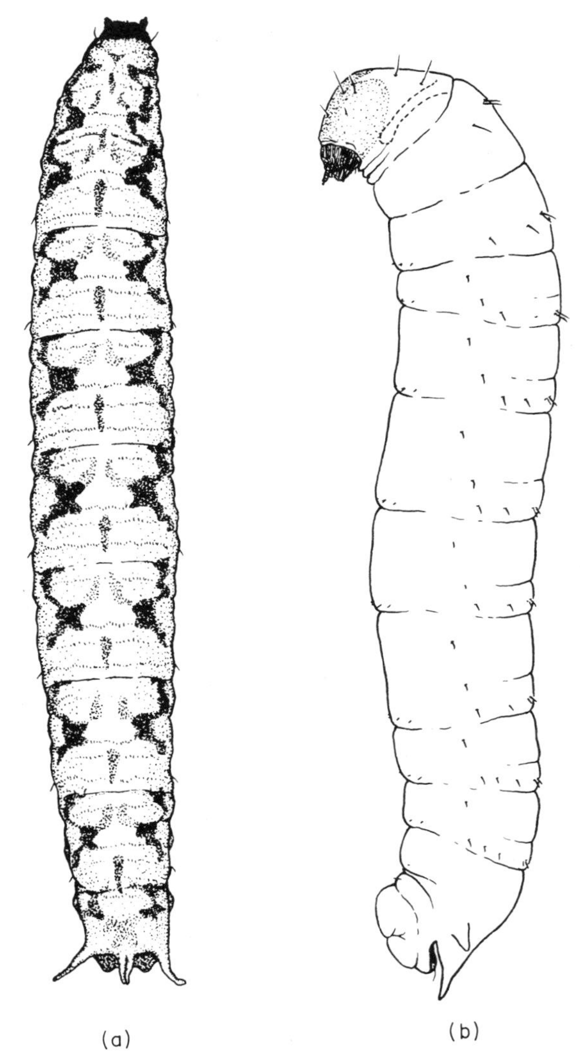

Fig. 7.3 Crane fly larvae of the genus *Dolichopeza* (a) subgenus *Dolichopeza* which is camouflaged by green and brown mottling and (b) subgenus *Oropeza* which is a uniform brown or green colour (Byers 1961).

expected to belong to this group. Among the biting midges, there are also a few species in the genus *Forcipomyia* which live in damp moss (Oldroyd 1964). Some Stratiomiid larvae also occur in wet moss cushions (Byers 1961).

3.4 Aphids

A number of aphids belonging to the genera *Myzodium* and *Muscaphis* have been found feeding on *Polytrichum* (Gerson 1969). More interesting are the aphids *Schlechtendalia chinensis*, *Nurudea shiraii* and *Nurudea rosea* which are responsible for producing gallnuts on sumach trees in the hilly region of Chekiang Province in China. Winged aphids arrive on the trees in spring, producing males and females viviparously. After mating, eggs are then laid in the young leaves resulting in the production of gallnuts which are of commercial importance. Three generations of aphids occur over the summer producing further nuts. The last is winged and fly off in search of the moss *Mnium* which is the secondary host. Here they give birth to 20–30 nymphs which become attached and feed on the tender young moss stalks secreting a waxy protective covering. After overwintering, the nymphs develop into the winged aphids which, in spring, fly off in search of sumach trees. Protection of the moss carpets and careful cultivation of the trees is recommended to improve gallnut production (Tang 1976). The galls contain 70–80% tannin and are used to produce black dyes for dyers and tanners as well as for ink. They are also employed medicinally as an expectorant and in the treatment of chancres and wounds (Porter-Smith & Wei 1969).

3.5 Mites

Mites are frequently found on mosses and one group of orabatid mites, the Cryptostigmata, are commonly called moss mites. Some of these, e.g. *Euphthiracarus flavum*, *Galumna nervosa*, *Oribotria* spp. and *Pseudotrita* spp. have been reared successfully on a diet made up exclusively of mosses. Their life cycle has been estimated to take from 50 to 70 days for its completion (Woodring 1963).

Another group of mites the Protostigmata contain the genus *Ledermuelleria* which has been isolated from mosses throughout the world. They are reddish in colour have reddish eggs and shiny black faecal droppings which facilitate their identification (Gerson 1969). These mites sit on the stems and leaves piercing the tissue with their sharp stylets and sucking out the cell contents of individual cells. As a result the young shoots lose their green colour, become silvery gray and may subsequently shrivel.

Almost half of more than 100 moss samples collected from Quebec, Canada, yielded mites. Species of *Ledermuelleria* were even found in

moss cushions beneath snow in midwinter. When brought into the laboratory, the mites commenced to oviposit onto suitable mosses, e.g. *Brachythecium*, *Hypnum*, *Didymodon* and *Ceratodon* (Gerson 1969). In addition to feeding on moss leaves there is a report of a mite that lays its eggs in the capsules of *Orthotrichum pusillum* (Keeley 1913).
There are also aquatic mites and some of these utilize mosses such as *Cratoneuron filicinum* for food. A study on the populations of water mites can enable the determination of water quality. For example in Belgium it was found that water polluted by industrial effluents caused a decrease in the density of water mites and change in the species composition. *Hygrobates fluviatilis*, the most pollution tolerant mite, increased in numbers whereas moss dwelling mites were less frequent as the water quality declined (Bolle *et al.* 1977).

3.6 Nematodes

These animals are frequently found in moss cushions and 15 different species have been isolated from such situations (Goodley 1963). As to whether they subsist directly on moss tissue or on organic debris trapped in the cushion has not been determined. Species such as *Plectus rhizophilus* are found in mosses from dry habitats whereas *Aphelenchoides parietinus* and *Plectus cirratus* are more typical of moist cushions which periodically dry. These nematodes can both swim and tolerate drought (Norton 1978).

Galls are frequently formed by plants in response to the presence of parasitic nematodes. The animals live and feed within the gall. Occasionally nematode induced galls are found on mosses and are hard grey or yellowish swollen bodies up to 2 mm long. They can be quite conspicuous with as many as fifty being found on a single stem (Dixon 1908; Sheldon 1936; Steiner 1936).

3.7 Slugs and snails

Damage is frequently observed to the leaves of a number of mosses which seem to be particularly attractive to these animals. Indeed, a fine set of specimens of *Buxbaumia aphylla* was destroyed by a slug which had managed to secrete itself in a parcel of moss being sent from Scotland to England (Stark 1860).

Moss feeding snails are not restricted to temperate regions. *Sphincterochila boissieri* and *Trochoidea seetzeni* are two species of snail which

Table 7.1 British birds which incorporate moss into their nests

Common name	Latin name	Common name	Latin name
Greylag goose	*Anser anser*	Robin	*Erithacus rubecula*
Whooper swan	*Cygnus cygnus*	Sedge warbler	*Acrocephalus schoenobaenus*
Red grouse	*Lagopus lagopus*		
Snipe	*Gallinago gallinago*	Dartford warbler	*Sylvia undata*
Whimbrel	*Numenius phaeopus*	Willow warbler	*Phylloscopus trochilus*
Kittiwake	*Rissa tridactyla*	Chiffchaff	*P. collybita*
Woodlark	*Lullula arborea*	Goldcrest	*Regulus regulus*
Great tit	*Parus major*	Firecrest	*R. ignicapillus*
Blue tit	*P. caeruleus*	Spotted flycatcher	*Muscicapa striata*
Coal tit	*P. ater*	Dunnock	*Prunella modularis*
Crested tit	*P. cristatus*	Meadow pipit	*Anthus pratensis*
Marsh tit	*P. palustris*	Tree pipit	*A. trivialis*
Willow tit	*P. montanus*	Pied/White wagtail	*Motacilla alba*
Long-tailed tit	*Aegithalos caudatus*	Grey wagtail	*M. cinerea*
Treecreeper	*Certhia familiaris*	Red-backed shrike	*Lanius collurio*
Short-toed treecreeper	*C. brachydactyla*	Hawfinch	*Coccothraustes coccothraustes*
Wren	*Troglodytes troglodytes*	Greenfinch	*Carduelis chloris*
		Goldfinch	*C. carduelis*
Dipper	*Cinclus cinclus*	Siskin	*C. spinus*
Mistle thrush	*Turdus viscivorus*	Linnet	*Acanthis cannabina*
Redwing	*T. iliacus*	Twite	*A. flavirostris*
Blackbird	*T. merula*	Redpoll	*A. flammea*
Wheatear	*Oenanthe oenanthe*	Crossbill	*Loxia curvirostra*
Stonechat	*Saxicola torquata*	Chaffinch	*Fringilla coelebs*
Whinchat	*S. rubetra*	Yellowhammer	*Emberiza citrinella*
Redstart	*Phoenicurus phoenicurus*	Snow bunting	*Plectrophenax nivalis*
		Tree sparrow	*Passer montanus*
Black redstart	*P. ochruros*		
Nightingale	*Luscinia megarhynchos*		

From Campbell and Ferguson-Lees (1972).

live together in the Negev Desert, Israel. The flora of the area consists of shrubs, lichens (e.g. *Ramalina maciformis* and *Buellia canescens*) and the moss *Desmatodon convolutus*. On the basis of plant remains in the digestive tract and excrement it is thought that *Sphincterochila boissieri*, a soil dweller, feeds on detritus, algae and lichens. *Trochoidea seetzeni*, which is mainly found in shrubs and feeds on them, also consumes *Desmatodon convolutus* and fruit bodies of *Buellia canescens* probably selecting the latter because they do not contain unpalatable lichen acids (Yom-Tov & Galun 1971).

4 PROTECTION AND CAMOUFLAGE

4.1 Protection

(a) *Aquatic mosses*

The larvae of some caddisflies hide within cases constructed from plant materials. The mosses *Fontinalis* and *Hygrohypnum* are used by *Adicrophleps hitchcockii* to make a log-cabin type of case which is square in cross-section. The larvae of blackflies (Diptera: Simuliidae) usually construct a papoose-like case in which the pupa develops. However many larvae that live among the leaves of *Fontinalis*, simply support the pupal case from the moss leaf blade by a few attaching threads (Glime 1978). The pupae hatch out in early summer into tiny biting flies that are a bane to field biologists in North America.

(b) *Terrestrial mosses*

Arthropods. The microclimate in moss cushions and carpets is much less extreme than that found in the environment immediately above. Thus in harsh climates, mosses are important for providing shelter for many different arthropods. In Antarctica, the more open cushions of *Polytrichum* and *Dicranum* harbour greater numbers of animals than the denser mats of *Pohlia* (Tilbrook 1967).

Birds. A wide range of birds use the green leafy gametophores of mosses to construct nests which provide protection for eggs and young. A list of some common British birds that incorporate mosses in their nests is given in Table 7.1 (Fig. 7.4). Sometimes nests are very complex, for example, the pink robin from Australia makes the main body of the nest from mosses (e.g. *Thuidium* and *Acrocladium*) and the lining from tree-fern down. The outside is ornamented with pieces of lichen kept in place with strands of spider's web (Newman & Bratt 1976).

Moss sporophytes are sometimes incorporated into the linings of birds nests. Thus a rufous-sided towhee from Virginia, U.S.A. interwove 70–80 capsule stalks to form an inner layer. Another unidentified bird from Michigan produced a very striking nest due to the golden colour imparted by the hundred or so stalks, with capsules attached, used in the lining. These were identified as belonging to the moss *Pohlia nutans* which has orange capsules and stalks (Crum 1973; Newman & Bratt 1976).

Fig. 7.4 The nest of a song thrush, *Turdus musicus*, one of many birds whose nest is constructed principally from mosses (photo: David Hosking).

Unfortunately, the species of mosses used by birds to build their nests are seldom mentioned in ornithology books even though some birds are prolific moss collectors. One nest of a phoebe from North America contained gametophores of *Mnium*, *Funaria*, *Polytrichum*, *Hypnum* and *Climacium*. A recent study has shown that carpet-forming (Pleurocarpous) mosses are most frequently used by birds to build their nests. A total of sixty different mosses were found by analysing the nests of eleven birds. It seems that in most cases, the moss species employed depends upon local availability, with the birds seldom showing preference for particular types (Breil & Moyle 1976).

4.2 Camouflage

The colouration of moss-dwelling caterpillars and larvae is often such that they can seldom be discerned with ease. However the most remarkable instance of insects being camouflaged against their natural background is found in the moss-forest weevils of New Guinea. These large leaf-eating weevils belonging to the genera *Gymnopholus* and *Pantorhites* live in the moist forests of the mountain ridges and summits. On their backs they possess patches which become colonized by algae, lichens, liverworts and mosses. The latter have been identified as *Daltonia angustifolia* and grow on five different species of weevils (Gressitt *et al.* 1968). The moss bearing species hardly carry any lichens, which are abundant on other weevils (Gressitt *et al.* 1965; Richardson 1975; Gerson & Seaward 1977). The backs of the weevils are structurally modified to provide niches for the growth of the plants. In the case of moss bearing species, numerous protonemata are found and mites which occur among the moss may transport moss propagules or spores from one host to another.

5 REFERENCES

Alexander C.P. (1920) The crane flies of New York. II. Biology and phylogeny. *Cornell Univ. Agric. Exp. Sta. Mem.* **38,** 691–1133.

Betts M.M. (1955) The food of titmice in an oak woodland. *J. Anim. Ecol.* **24,** 282–323

Bolle D., Wauthy G. & Lebrun P. (1977) Preliminary studies of watermites (Acari. Prostigmata) as bioindicators of pollution in streams. *Ann. Soc. R. Zool. Belg.* **106,** 201–10.

Breil D.A. & Moyle S.M. (1976) Bryophytes used in construction of bird nests. *Bryologist,* **79,** 95–8.

Brindle A. (1964) Taxonomic notes on the larvae of British Diptera. No. 18—The Memerodrominae (Empididae). *Entomologist,* **97,** 162–5.

Byers G.W. (1961) The crane fly genus Dolichopeza in North America. *Univ. Kansas Sci. Bull.* **42,** 665–924.

Campbell B. & Ferguson-Lees J. (1972) *A Field Guide to Bird's Nests.* Constable, London.

Chapman T.A. (1894) Some notes on micro-lepidoptera whose larvae are external feeders and chiefly on the early stages of *Eriocephala calthella* (Zygaenidae, Lymacodidae, Eriocephalidae). *Trans. R. Entomol. Soc. Lond.* **1894,** 335–50.

Clarke G.C.S., Greene S.W. & Greene D.M. (1971) Productivity of bryophytes in polar regions. *Ann. Bot. (Lond.),* **35,** 99–108.

Collins N.J. & Oechel W.C. (1974) The pattern of growth and translocation of photosynthate in the tundra moss *Polytrichum alpinum. Can. J. Bot.* **52,** 355–63.

Corbet S.A. & Lan O.B. (1974) Moss on a roof and what lives in it. *J. Biol. Educ.* **8,** 153–60.

CRUM H. (1973) Mosses of the Great Lakes Forest. *Contrib. Univ. Mich. Herb.* **10,** 1–404.
DICKINSON C.H. & MAGGS G.H. (1974) Aspects of the decomposition of *Sphagnum* leaves in an ombrophilous mire. *New Phytol.* **73,** 1249–57.
DIXON H.N. (1908) Nematode galls on mosses. *Bryologist,* **11,** 31.
GARDARSSON A. & SIGURDSSON J.B. (1972) *Research on the Pink-footed Goose.* Mimeographed Report, Icelandic National Energy Authority. Reykjavik.
GERSON U. (1969) Moss–Arthropod associations. *Bryologist,* **72,** 495–500.
GERSON U. & SEAWARD M.R.D. (1977) Lichen–invertebrate associations. In *Lichen Ecology,* ed. Seaward M.R.D., pp. 69–119. Academic Press, London.
GLIME J.M. (1978) Insect utilization of bryophytes. *Bryologist,* **81,** 186–7.
GOODLEY J.B. (1963) *Soil and Fresh Water Nematodes,* 2nd Ed. Methuen, London.
GRESSITT J.L., SEDLACEK J. & SZENT-IVANY J.J.H. (1965) Flora and fauna on the backs of large Papuan Moss-Forest Weevils. *Science* (*Wash. D.C.*), **150,** 1833–5.
GRESSITT J.L., SAMUELSON G.A. & VITT D.H. (1968) Moss growing on living Papuan moss-forest weevils. *Nature* (*Lond.*), **217,** 765–7.
HANSSON L. (1979) Condition and diet in relation to habitat in bank voles *Clethrionomys glareolus:* population or community approach. *Oikos,* **33,** 55–63.
JONES J.R.E. (1949) A further ecological study of calcareous streams in the 'Black Mountains' district of South Wales. *J. Anim. Ecol.* **18,** 142–59.
KALLIO P. & KARENLAMPI L. (1975) Photosynthesis in mosses and lichens. In *Photosynthesis and Productivity in Different Environments,* ed. Cooper J.P., pp. 393–423. Cambridge University Press.
KAUFMANN T. (1965) Biological studies on some Bavarian Acridoides (Orthoptera) with special reference to their feeding habit. *Ann. Entomol. Soc. Am.* **58,** 791–801.
KEELEY F.J. (1913) Eggs of a mite in empty capsules of *Orthotrichum pusillum. Bryologist,* **16,** 18–19.
LID J. & MEIDELL O. (1933) The food of Norwegian grouse chicks. *Nytt. Mag. Naturvidensk,* **73,** 75–114.
MARGARIS N.S. & KALAITZAKIS J. (1974) Soluble sugars from some Greek mosses. *Bryologist,* **77,** 470–2.
NEWMAN O.M.G. & BRATT G.C. (1976) The Pink Robin—an avian lichen buff. *Search* (*Syd.*), **7,** 487–8.
487–8.
NORTON D.C. (1978) Ecology of plant-parasitic nematodes, pp. 51–7. John Wiley, New York.
OECHEL W.C. & SVEINBJORNSSON B. (1978) Primary production processes in arctic bryophytes at Barrow, Alaska. *Ecol. Stud.* **29,** 269–98.
OLDROYD H. (1964) *The Natural History of Flies.* Weidenfeld & Nicolson, London.
PAKARINEN P. & VITT D.H. (1974) The major organic components and calorie contents of high arctic bryophytes. *Can. J. Bot.* **52,** 1151–61.
PENNY N.D. (1977) A systematic study of the family Boreidae (Mercoptera). *Univ. Kansas Sci. Bull.* **51,** 141–217.
PORTER-SMITH F. & WEI P.D. (1969) *Chinese Materia Medica: Vegetable Kingdom,* p. 182. Kuting Book House, Taipei, Taiwan.
RAGGE D. (1965) *Grasshoppers, Crickets and Cockroaches.* F. Warne, London.
REIMERS E. & KLEIN D.R. (1979) Reindeer and Caribou. *Nature* (*Lond.*), **282,** 558–9.
RICHARDSON D.H.S. (1975) *The Vanishing Lichens.* David & Charles, Newton Abbot.
RUSSELL L.K. (1979) A new genus and a new species of Boreidae from Oregon (Mercoptera). *Proc. Ent. Soc. Washington,* **81,** 22–31.
SHELDON J.L. (1936) Nematode galls in bryophytes. *Bryologist,* **39,** 94–5.
SHORTHOUSE J.D. (1979) Observations of the snow scorpionfly *Boreus brumalis* Fitch (Boreidae: Mercoptera) in Sudbury, Ontario. *Quaestiones Entomologicae,* **15,** 341–4.

STARK R.M. (1860) *A Popular History of British Mosses*, p. 8, George Routledge, London

STEERE W.C. (1974) The status and geographical distribution of *Voitia hyperborea* in North America (Musci: Splachnaceae). *Bull. Torrey Bot. Club*, **101,** 55–63.

STEINER G. (1936) *Anguillulina askenasyi* (Bütschli 1873), a gall forming nematode parasite of the common fern moss, *Thuidium delicatulum* (L.) *Hedw. J. Wash. Acad. Sci.* **26,** 410–14.

SUGAWA (1960) Nutritive value of mosses as food for domestic animals and fowls. *Hikobia,* **2,** 119–24.

TANG C. (1976) The Chinese gallnuts, their multiplication and means for increasing production. *Acta Entomol. Sin.* **19,** 282–96.

TILBROOK P.L. (1967) Arthropod ecology in the maritime Antarctica. In *Entomology of Antarctica*, ed. Gressitt J.L., pp. 331–56. Washington, D.C.

UKRAINTSEVA V.V., FLEROV K.K. & SOLONEVICH N.G. (1978) Analysis of plant remains from the alimentary tract of Mylakhchinsk bison (Yakutia). *Bot. Zh. S.S.S.R.*, **63,** 1001–4.

UVAROV B. (1977) *Grasshoppers and Locusts*. Vol. 2, p. 103. Centre for Overseas Pest Research.

VERDCOURT B. (1947) A note on the food of *Acrydium* Geoff. (Orthopt.). *Entomol. Mon. Mag.* **83,** 190.

WOODRING J.P. (1963) The nutrition and biology of saprophytic sarcoptiformes. *Adv. Acarol.* **1,** 89–111.

YOM-TOV Y. & GALUN M. (1971) Notes on the feeding habits of the desert snails *Sphinctero-chila boissieri* Charpentier and *Trochoidea* (*Xerocrassa*) *seetzeni* Charpentier. *Veliger* (*Quart. Calif. Malacazool. Soc.*), **14,** 86–9.

CHAPTER 8
MOSSES AND MICRO-ORGANISMS

1 BACTERIA

1.1 Decomposition by bacteria

Bacteria do not seem to be very important in the decomposition of mosses, at least in *Sphagnum* mires. A few actinomyctes are found associated with moss leaves in this habitat, and it seems likely that bacteria only become important at depths in the peat where oxygen is absent and anaerobic species can thrive though the rate of decomposition is very slow (Dickinson & Maggs 1974). Even in more favourable situations breakdown of moss tissue is not rapid and this may be due to the production of antibiotics by mosses.

Antibiotics active against at least one bacterium have been found in extracts of 55% of the mosses investigated (61 out of 111 species). The genera which seem most active in antibiotic production are *Atrichum*, *Cratoneuron*, *Dicranum*, *Polytrichum* and *Sphagnum*. *Brachythecium* has a particularly broad spectrum of activity, inhibiting two gram positive and four gram negative bacteria as well as the acid fast *Mycobacterium*. Extracts of *Funaria hygrometrica* had no effect against any test organism. Perhaps the reason why this moss has not evolved the capacity for antibiotic synthesis is related to the fact that it quickly exploits open habitats following fires. In such situations, there is likely to be little competition from other organisms. The compounds responsible for the observed antibiotic activity in mosses have not been isolated (Banerjee & Sen 1979).

1.2 Bacterial–moss associations

Agrobacterium tumefaciens causes crown gall, a bacterial disease which reduces the vigour of attacked higher plants and gives rise to unsightly galls through which other pathogenic micro-organisms can gain entry. In one state of the U.S.A., California, the disease is estimated to cost nursery and orchard owners 7 million dollars annually (Lippincott & Lippincott 1975). Research into *Agrobacterium tumefaciens* has been

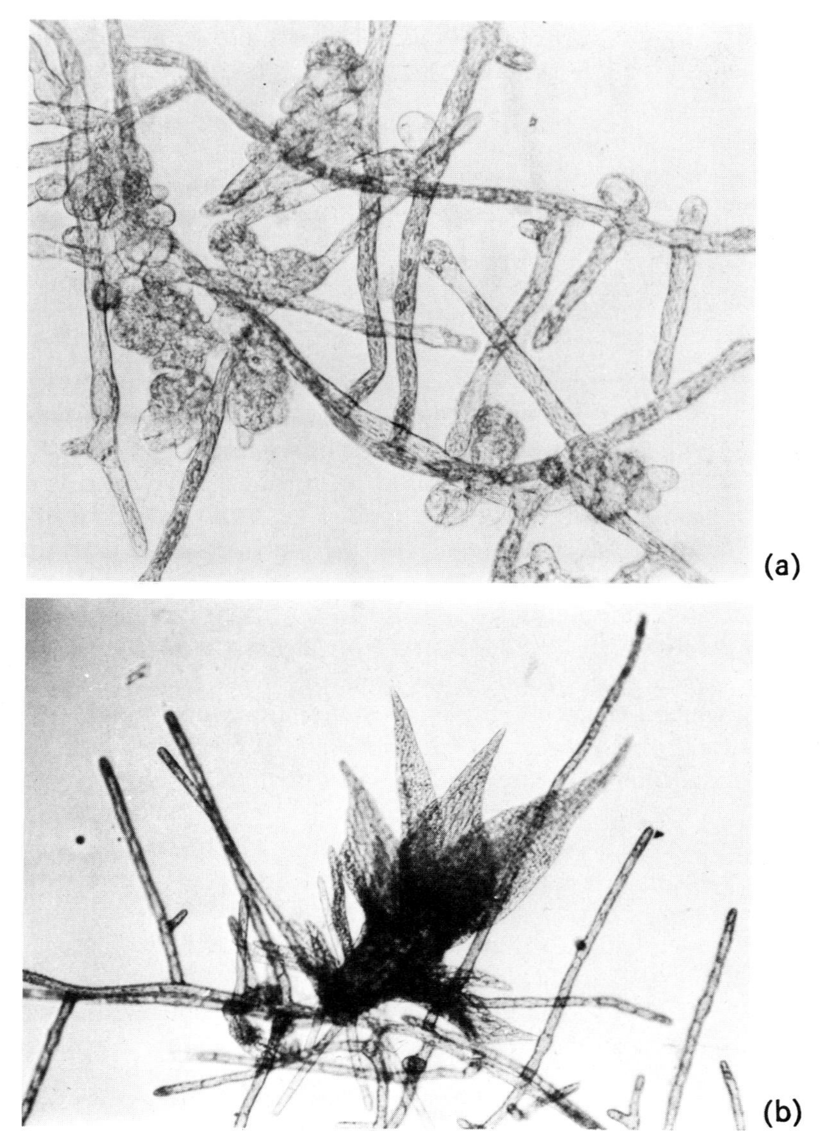

Fig. 8.1 (a) Clusters of gametophore buds induced on the protonema of *Pylaisiella selwynii* by *Agrobacterium tumefaciens*. (b) A normal protonema of this moss with a developing bud (Spiess *et al.* 1971).

advanced by the discovery that the bacterium can form an association with the moss *Pylaisiella selwynii*. In flowering plants, *Agrobacterium tumefaciens* is normally only able to gain entry via a wound. A tumour-inducing principle is then formed. This causes the host cell to synthesize plant hormones and other substances, to multiply and then to be no longer under control. A tumour results and it is thought that the tumour-inducing principle, a plasmid, involves bacterial genes becoming incorporated into the host cell. The bacteria themselves are also able to produce the plant hormones auxin, cytokinin and giberellin and the former at least seems to enhance conversion of host cells to tumour forming cells.

Studies on the moss mentioned above have enabled understanding of tumour induction. In this moss, the bacterium attacks the protonema and causes an increased production of gametophore buds rather than tumours (Fig. 8.1) (Spiess *et al.* 1971). By separating the bacteria and moss with filters (which allow the passage of plant hormones but not bacteria) it has been shown that actual contact of the bacteria with the moss is essential before buds will be induced. This feature is in common with flowering plant infections (Spiess *et al.* 1976). Recent studies have shown that only a combination of cytokinin (0·01 μM) and octopine (0·1 mM) added to cultures of the moss results in enhanced bud production comparable to that induced by *Agrobacterium* infections. Octopine is an amino acid which is produced by crown-gall bacteria and is evidently important in the differentiation of buds in the moss *Pylaisiella* (Lippincott & Lippincott 1979). The bacteria bind at specific sites on the moss protonema by means of a special lipopolysaccharide (Whatley & Spiess 1977). *Agrobacterium*-like bacteria developed in cultures made from washings of *Pylaisiella selwynii* collected from its natural habitat on decaying trunks and logs. This suggests that the moss-bacterial association is a natural one and normally plays a role in the development of this moss (Spiess & Lippincott 1978). The moss-*Agrobacterium* system is an ideal one for research to help understand crown gall disease in higher plants because little space is necessary for experiments, wounding is not necessary and results can be obtained by examining moss cultures only three weeks after introduction of the bacteria.

2 PROTOZOA

Protozoa are tiny single celled animals and little is known about their relationships with mosses except in the case of *Sphagnum*. The structure of this moss furnishes an aquatic habitat whose fauna is unlike that found elsewhere. Nearly all the organisms are small enough to inhabit the film of water in the concavity of the *Sphagnum* leaf which may be only 300 μm in diameter (Corbet 1973). One investigation in North America revealed 20 species of amoebae (Sarcodina), 20 species of flagellates (Mastigophora), and 6 species of ciliates (Ciliophora) in the water in or on the *Sphagnum* plants. The protozoa near the top of the plant generally possess chlorophyll or contain symbiotic green algae, while colourless forms are found lower down (Chacharonis 1956).

The protozoan fauna of *Sphagnum* is dominated by shelled amoebae (testate rhizopods) (Fig. 8.2). These little animals construct shells,

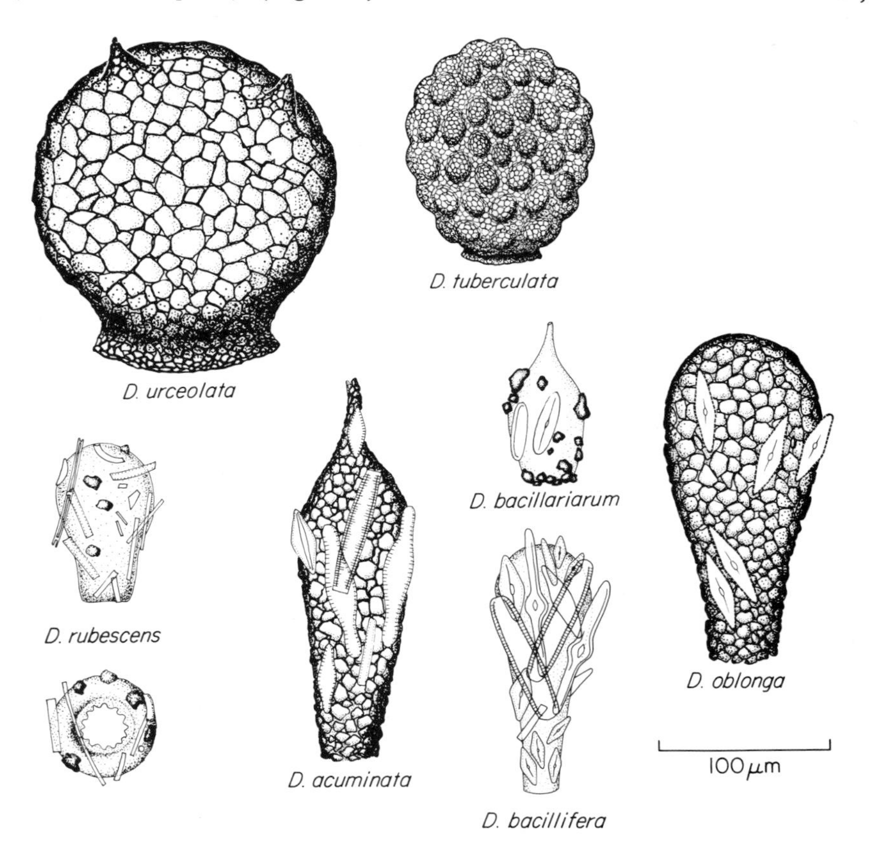

Fig. 8.2 A range of ***Difflugia*** **spp., a genus of shelled amoebae, which live on the surface of *Sphagnum* leaves. The amoebae live within the shell and put out finger-like extensions of protoplasm from the base which pick up food from the leaf surface. (Corbet 1973).**

known as tests, from mineral and organic particles which they pick up from their environment. The tests persist long after the animals have died. These amoebae feed on algal cells (growing epiphytically on the *Sphagnum* leaves) on each other and also on bacteria in the case of the smaller amoebae. As mentioned above a vertical zonation of species is found on the moss. Thus of three species found in England which contained green algal symbionts, two occurred in the upper 6 cm of the moss. The other was found from 4–10 cm which possibly represents a balance between its requirement for light and for peat and mineral particles from which to construct its shell. All but two of the 27 species lacking symbionts showed maximum abundance below 6 cm in the *Sphagnum*. Unlike other protozoa, which tend to move down to moister situations when conditions become unfavourable, the shelled amoebae form little cysts and survive dry periods by this means (Heal 1962).

It has been estimated that there may be as many as 16 million living and 20 million shells of dead amoebae per m^2 of a *Sphagnum* carpet of depth 12 cm. This compares with 8 million living and 25 million dead amoebae of this type in the top 40 cm of soil in an oak wood. Thus shelled amoebae are an abundant and important part of the fauna of *Sphagnum* bogs (Chiba & Kato 1969).

3 FUNGI

3.1 FILAMENTOUS TYPES

The range of fungi which parasitize mosses is quite broad, though only recently has a systematic survey been attempted and almost no experimental work has been done (Doebbeler 1978). Some, e.g. *Pleotrachelus wildermanii*, one of the Mastigomycotina (lower fungi), attack the rhizoids of *Funaria hygrometrica*, and probably other mosses, causing swelling of the host cells and depletion of starch reserves in the host (Fig. 8.3; Ingold 1952). Other parasitic fungi belong to the Ascomycotina (cup fungi) and attack the gametophores. A few of these are known to parasitize mosses (Prior 1966; Bowen 1968). There is also an interesting jelly fungus related to rusts and belonging to the Basidiomycotina which is found on a wide range of mosses including *Brachythecium*, *Climacium*, *Hypnum*, *Plagiothecium* and *Thuidium*. The fungus *Eocronartium muscicola* belongs to the fungal order Auriculariales. It invades the moss host and persists from year to year but in spring grows

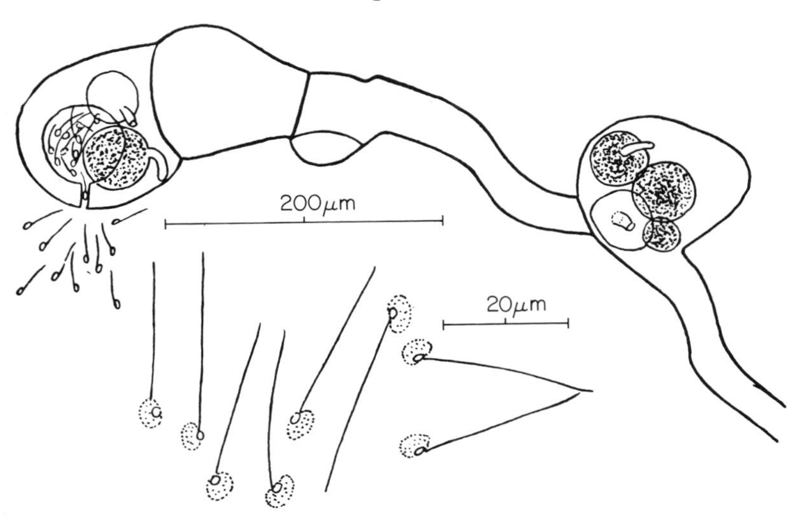

Fig. 8.3 A rhizoid of *Funaria hygrometrica* attacked by the fungus *Pleotrachelus wildermanii*. Above, part of a rhizoid of *Funaria* with two infected cells: one sporangium of the fungus is liberating its zoospores. Below, single zoospores more highly magnified (Ingold 1952).

Fig. 8.4 *Climaceum americanum* infected by the jelly fungus *Eocronartium muscicola.* Note the fruit bodies of this systemic infection growing from the apices of the moss shoots (Fitzpatrick 1918).

up through the stems and leaves of the young developing gametophore, enveloping the growing apex. The fungus then emerges and filaments coalesce to form a club-shaped fruit body which may be 2–6 cm high looking rather like a malformed moss capsule (Fig. 8.4). Dozens of these club-shaped bodies, one from the apex of each moss shoot, may be seen on an infected moss tuft (Fitzpatrick 1918). Basidiospores are formed on the surface of the fruit bodies and shot off or splashed off by rain. Presumably on landing on the apex of a moist moss shoot they are able to initiate new infections.

Parasitism of mosses by fungi is occasionally recorded in temperate zones but seems to be more frequent in colder regions. This may only be because damage to mosses is more obvious in regions where there is little other vegetation. Circular patches of fungal infection occur on *Rhacomitrium canescens* in the arctic and on other mosses, particularly from wet communities in the antarctic. Thus moss carpets of *Brachythecium* and *Drepanocladus* as well as moss cushions of wet ledges formed by species of *Ceratodon* and *Dicranoweisia* frequently show signs of attack. The symptoms are whitish patches or rings on the moss (Fig. 8.5) marking the position of fungal infection on the moribund shoots. The fungus grows radially and the diameter of the infected area

Fig. 8.5 Moss cushions from the maritime Antarctic with radial patterns of infection by the fungus. *Thyronectria antarctica* (photo: R.E. Longton).

can increase by an estimated 25 mm during the 4 month growing period. With the exception of *Dicranoweisia*, parasitized mosses are not killed and eventually recover and regenerate. At least two fungi are responsible for the observed damage and one, *Thyronectria antarctica*, is a member of the Ascomycotina (cup fungi) mentioned earlier (Longton 1973).

Filamentous fungi are also important in the decomposition of dying moss leaves. In *Sphagnum* bogs, *Mortierella* and *Cladosporium* consume the cell contents of senescing leaves. Other fungi such as *Penicillium*, *Geotrichum* and *Tolypocladium* (Fungi Imperfecti) as well as various Basidiomycotina, then partially break down the cell wall tissue of the *Sphagnum* cells releasing soluble nitrogenous compounds. Most of this decomposition takes place in the top 15 cm of the *Sphagnum* carpet. Below this the acid condition and low oxygen levels result in the death of the fungi (and other decomposers) so that little further breakdown occurs and the peat deposit progressively accumulates (Dickinson & Maggs 1974).

3.2 Yeast-like forms

From December 1975 until February 1976 there was an epidemic in Mississippi, U.S.A. of cutaneous sporotrichosis among workers picking pine seedlings. Sporotrichosis is an occupational disease of horticulturists especially in the warmer humid regions of the world and its incidence is frequently associated with *Sphagnum* (Kedes *et al.* 1964; Rippon 1974).

The disease is caused by a fungus, *Sporothrix schenckii* and if spores of the fungus get into a wound they lead to the following symptoms: an ulcer develops at the site of infection with a series of sucutaneous nodules and inflammation in the general area (Fig. 8.6), while occasional ulceration may occur along the course of the lymphatic drainage system. The ulcers do not respond to antibiotics or to surgical drainage but are normally cleared up by oral administration of potassium iodide, though how this effects a cure is not understood.

In the Mississippi outbreak the cause of the disease was traced to contaminated *Sphagnum* moss at one nursery. Investigation showed that this batch of *Sphagnum* had been stored outdoors rather than inside (the normal practice) because all the store rooms were full. It is believed that the moss was lightly contaminated when harvested from the peat bog or, became so shortly afterwards. (The fungus can live on dead organic

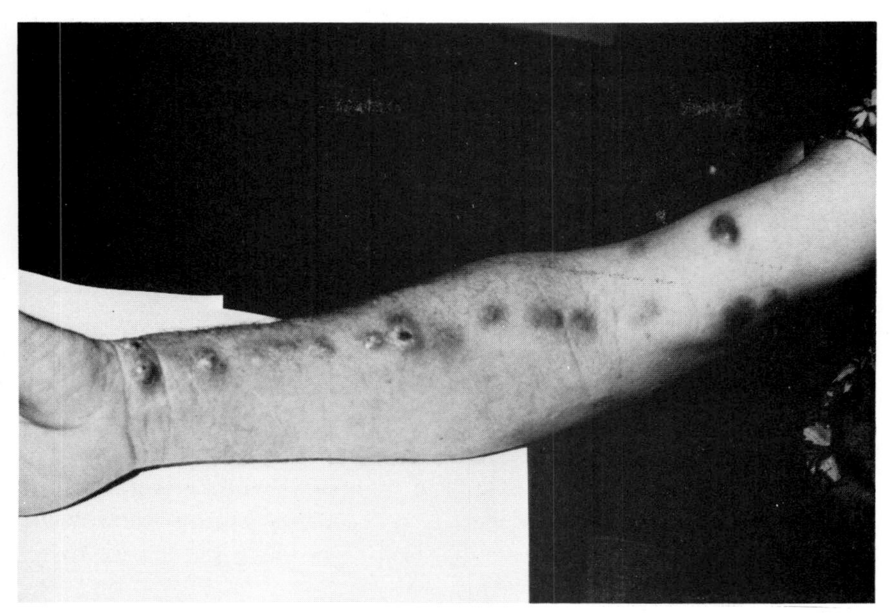

Fig. 8.6 A case of sporotrichosis contracted by a horticultural worker from contaminated *Sphagnum* (Powell *et al.* 1978).

material, including moss, as well as causing a disease in humans.) It seems that storage of the moss outdoors in the moist warm Mississippi summer provided ideal conditions for the fungus so that by the time the moss was used for packing pine seedlings in the winter it was heavily contaminated (Powell *et al.* 1978). It is interesting to note that in a recent survey antibiotics effective against filamentous fungi were not found to be produced by mosses (Banerjee & Sen 1979). However, compounds effective against the yeastlike *Candida albicans*, a relative of *Sporothrix*, have been discovered in *Sphagnum portoricense* and two other mosses (Madsen & Pates 1952; McCleary *et al.* 1960). Possibly the low rate of infection by *Sporothrix schenckii* in healthy *Sphagnum* collected from the field may be related to the production of such anti-fungal substances by the moss.

The preventive measures which were introduced after the epidemic of sporotrichosis included burying remaining stocks of contaminated moss under 60 cm of soil; storing all moss indoors; monthly scrubbing of storage and packing buildings with disinfectant; reduction in contact between the moss and skin of workers; and testing samples of moss regularly for the presence of *Sporothrix schenckii*. These precautions cost the Mississippi Forestry Commission about 3,000 dollars per year (Powell *et al.* 1978). *Sphagnum* also seems to be a resevoir for Mycobacteria in regions of Norway with a former high incidence of leprosy. The bacteria can be cultured by foot-pad innoculation into mice, an indication that the species may be *Mycobacterium leprae* (Kazda *et al.* 1980).

4 ALGAE

4.1 BLUE-GREEN ALGAE

The structure and reproduction of these algae is more akin to bacteria so that they are now frequently referred to as the Photosynthetic Cyanobacteria. Although the individual filaments or cells are microscopic, the gelatinous colonies formed by many species are often quite large and easily seen with the naked eye or a lens. Unlike other algae, but similar to some bacteria, blue-green algae have the ability to fix nitrogen from the air and incorporate it into amino acids and proteins. In nutrient poor habitats this capability is important for surrounding organisms as part of the fixed nitrogen is released and the remainder becomes available on the death of the algae, so enriching its environment. Many different organisms, including a few mosses, have taken advantage of this ability to fix nitrogen by forming associations with blue-green algae.

(a) *True mosses*

The island of Surtsey was formed off the coast of Iceland in 1963 as a result of an under-sea volcanic eruption. Biologists have studied the colonization of the island (area of about three square kilometres) by various plants and animals. Mosses and algae were among the early arrivals. A survey 6 years after the island's formation showed that wherever *Funaria hygrometrica* was found the soil around and below the moss was colonized by blue-green algae, especially *Anabaena variabilis*. It was shown experimentally that the algae release substantial amounts of fixed nitrogen which is taken up by the moss. In addition, the number of gametophores was greater in moss cultures on sand which had been inoculated with the alga than on controls. Apparently the algae produce growth substances which benefit the moss. In turn *Funaria hygrometrica* has a growth-promoting effect on the algae. This association may therefore be important for the colonization of the island both by enhancing the establishment of moss and algae and indirectly by increasing the organic matter in the soil (through their growth and decomposition). This reduces soil erosion, and improves the retention of soil moisture so that other plants may thrive (Rodgers & Henriksson 1976).

Association between blue-green algae and mosses occur in other situations. Blue-green algae growing on the surface of *Drepanocladus* in pools on the edge of subarctic Swedish mires fix atmospheric nitrogen which is thought to benefit both the moss and the surrounding plants (Basilier *et al.* 1978). Nitrogen fixed by algae growing in the top centimetre of soils in sand dune slacks has been shown to move rapidly to mosses such as *Bryum pendulum* (as well as to higher plants) growing in the same habitat (Stewart 1967). Similarly nitrogen fixed by the blue-green alga *Nostoc* growing near *Gymnostomum recurvirostre* is eventually taken up by the moss (Jones & Wilson 1978).

(b) *Peat mosses*

The large colourless dead cells of *Sphagnum* leaves can be invaded by the blue-green alga *Hapalosiphon* which may then become imprisoned in the cells. Nitrogen fixed by the alga may pass to the moss (Stewart 1966). Of greater practical importance are the colonies of blue-green algae that live in quite large numbers on the surface of *Sphagnum* plants.

In the mires of Swedish Lapland blue-green algae, which grow epiphytically on mosses, are the main or only organisms capable of fixing nitrogen, and hence play an important role in this nitrogen-poor ecosystem. It is estimated that they contribute between 0·5–6·4 g of nitrogen/m^2/yr with an average of about 3 g/m^2/yr. Fixation is optimal in summer when the temperature of the bogs is about 16°C. The *Sphagnum* plants on the edge of the mire (through the activity of their associated algae) show fixation rates some six times greater than plants in the centre of the mire. This is due to more algae being present on the former probably because the opportunities for colonization of the moss are greater there. In the centre of the mire growth on all but the upper portions of the moss, and movement between plants, may be inhibited by the acid anaerobic conditions. No correlation was observed between nitrogen fixation rates and the growth of individual moss plants. This is not surprising since much of the fixed nitrogen is probably exuded and absorbed by several moss shoots. Other organisms, especially on the mire margin, where fixation rates are highest, are likely to benefit from nitrogen fixed by the algal–moss association. This is indicated by the abundant dark green shoots of cotton grass growing in such areas (Basilier *et al.* 1978).

Highly active nitrogen fixing moss–algal associations are also known from the arctic and subarctic mires of Alaska and Finland as well as from mires from more temperate regions around Uppsala in southern Sweden. The alga–*Sphagnum* association may play an important role in other vegetation types. For example, wet *Sphagnum* communities cover 17% of the ground in the pine-spruce forests of central Sweden and the algae associated with the moss can fix 0·8–3·8 g nitrogen/m^2/yr (Basilier *et al.* 1978; Granhall & Lindberg 1978).

4.2 Other algae

Many different sorts of algae live attached to mosses, especially those which grow in moist habitats. On Signy Island, located off the coast of Antarctica, algae are more important associates of the two dominant mosses than either bacteria or fungi. The algal counts on a single *Polytrichum alpestre* shoot amounted to approximately 12×10^6 algal cells per gram of moss. Most of the algae occur as colonies of from 4–10 cells attached to the surface of the leaves and over 90% of the algae are found within 2 mm of the leaf base. The number of algae on the youngest leaf is low and increases successively to the seventh or eighth leaf and then decreases probably due to shading. Of eleven species of algae identified, *Monodus subterraneous* (a yellow-green alga) and two unknown species of Radiococcaceae were most frequent, with *Chloromonas* and the filamentous *Gongriosira* (all green algae) also being common. Blue-green algae were few on the mosses on Signy Island, possibly because conditions were too acid. Algae in the mosses on the island play an important role as a source of food for other organisms. The mucilage produced by the colonies may be used by bacteria, yeasts and other fungi. These in turn may be consumed by protozoa and springtails (Broady 1977).

In *Sphagnum* carpets, unicellular desmids, which float freely rather than being attached, are very abundant in water trapped in or on the moss carpet. Diatoms are also frequent in this habitat and parts of their shells may be incorporated into the tests of the shelled amoebae mentioned earlier.

A few green algae, particularly those of the Chlorococcales form more intimate associations with a variety of plants. These algae grow either inter- or intra-cellularly and are termed space parasites. One occurs on *Sphagnum* plants and its presence is identified by irregular vivid green blotches on old bleached leaves. The bright colour is due to the resting spores of the alga named *Phyllobium sphagnicola*. The spores are thought to release motile cells (zoospores) which settle on the surface of a leaf penetrating the large dead cells possibly via a pore. Once inside, a green filament is formed which grows through the leaf occasionally forming gametangia; these release motile gametes, fuse in pairs, and invade new hosts. Eventually the green filaments produce a mass of the resting cells on the surface and edge of the leaf and the filaments at maturity lose their chlorophyll (Fig. 8.7). The moss does not appear to be much damaged by the presence of the alga though this has not been shown experimentally (West 1908; Fritsch 1935).

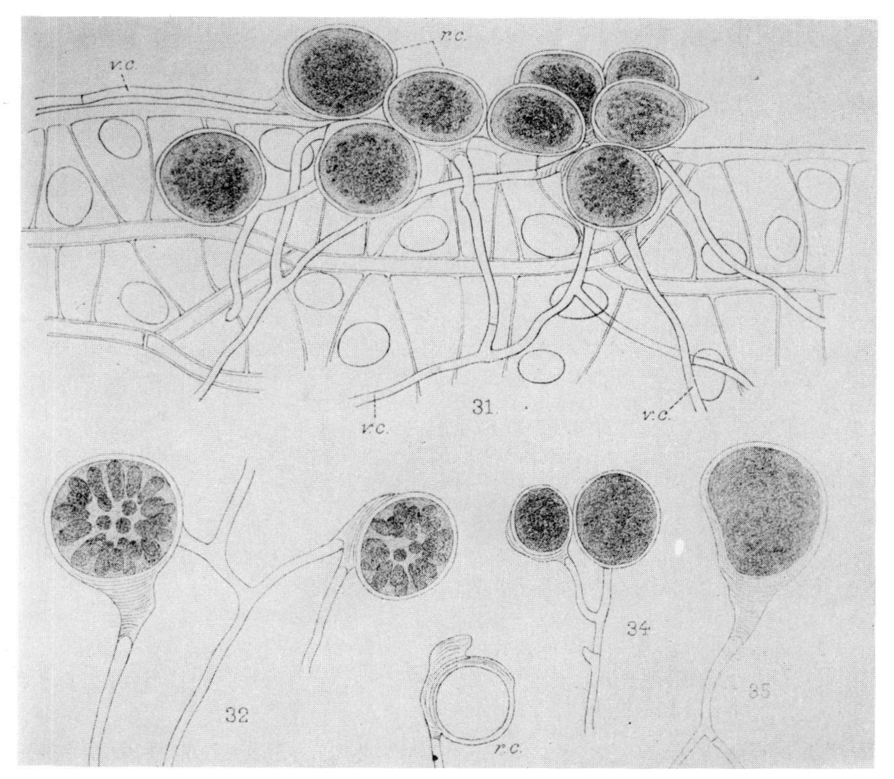

Fig. 8.7 *Sphagnum* cells invaded by the green alga *Phyllobium sphagnicola*, (31) edge of *Sphagnum* leaf with resting cells attached to ramifying vegetative filament, (32–35) details of resting cells (West 1908).

5 REFERENCES

Banerjee R.D. & Sen S.P. (1979) Antibiotic activity of bryophytes. *Brologist* , **82,** 141–53.

Basilier K., Granhall U. & Stenstrom T.A. (1978) Nitrogen fixation in wet minerotrophic moss communities of a subarctic mire. *Oikos,* **31,** 236–46.

Bowen W.R. (1968) The imperfect fungus *Schizotrichella lunata* on the moss *Mnium cuspidatum. Bryologist,* **71,** 124–6.

Broady P.A. (1977) The Signy Island terrestrial reference sites, VII. The ecology of the algae of site 1, a moss turf. *Br. Antarct. Surv. Bull.* **45,** 47–62.

Chacharonis P. (1956) Observations on the ecology of protozoa associated with *Sphagnum. J. Protozool.* **3** (*Suppl.*), abstr. 58.

Chiba Y. & Kato M. (1969) Testacean community in the bryophytes collected in the Mt. Kurikoma District. *Ecol. Rev.* **17,** 123–30.

Corbet S.A. (1973) An illustrated introduction to the testate rhizopods in *Sphagnum,* with special reference to the area around Malham Tarn, Yorshire, *Field Stud,* **3,** 801–38.

Dickinson C.H. & Maggs G.H. (1974) Aspects of the decomposition of *Sphagnum* leaves in an ombrophilous mire. *New Phytol.* 1249–57.

Doebbeler P. (1978) Ascomycete parasites of moss. 1. Pyrenomycete species occurring on gametophytes. *Mitt. Bot. Staatsamml. Muench.* **14,** 1–360.

Fitzpatrick H.M. (1918) The life history and parasitism of *Eocronartium muscicola. Phytopathology,* **8,** 197–218.

Fritsch F.E. (1935) *The Structure and Reproduction of the Algae, vol. I,* pp. 154–8. Cambridge University Press.

Granhall U. & Lindberg T. (1978) Nitrogen fixation in some coniferous forest ecosystems. *Ecol. Bull.* **26,** 178–92.

Heal O.W. (1962) The abundance and micro-distribution of testate amoebae (Rhizopoda: Testacea) in *Sphagnum. Oikos,* **13,** 35–47.

Ingold C.T. (1952) *Funaria* rhizoids infected with *Pleotrachelus wildermanii, Trans. Br. Bryol. Soc.* **12,** 53–4.

Jones K. & Wilson R.E. (1978) The fate of nitrogen fixed by a free-living blue-green alga. *Ecol. Bull.* **26,** 158–63.

Kazda J., Irgens, L.M. & Muller K. (1980) Isolation of non-cultivable acid-fast bacilli in *Sphagnum* and moss vegetation by foot pad technique in mice. *Int. J. Leprosy,* **48,** 1–6.

Kedes L.H., Siemienski J. & Braude A.I. (1964) The syndrome of the alcoholic rose gardener. *Ann. Intern. Med.* **61,** 1139–41.

Lippincott J.A. & Lippincott B.B. (1975) The genus *Agrobacterium* and plant tumorigenesis. *Annu. Rev. Microbiol.* **29,** 377–405.

Lippincott B.B. & Lippincott J.A. (1979) Octopine–cytokinin control of moss development. *Abstracts Xth Int. Conf. Plant Growth Substances,* p. 33. Madison, Wisconsin.

Longton R.E. (1973) The occurrence of radial infection patterns in colonies of polar bryophytes. *Br. Antarc. Surv. Bull.* **32,** 41–9.

Madsen G.C. & Pates A.L. (1952) Occurrence of antimicrobial substances in chlorophyllose plants growing in Florida. *Bot. Gaz.* **113,** 293–300.

McCleary J.A., Sypherd P.S. & Walkington D.L. (1960) Mosses as a possible source of antibiotics. *Science* (*Wash. D.C.*), **131,** 108.

Powell K.E., Taylor A., Phillips B.J., Blakey D.L., Campbell G.D., Kaufman L. & Kaplan W. (1978) Cutaneous sporotrichosis in forestry workers. *J. Am. Med. Assoc.* **240,** 232–5.

Prior P.V. (1966) A new fungal parasite of mosses. *Bryologist,* **69,** 243–6.

Rippon J.W. (1974) *Medical Mycology,* p. 250. Saunders, Philadelphia.

RODGERS G.A. & HENRIKSSON E. (1976) Associations between the blue-green algae *Anabaena variabilis* and *Nostoc muscorum* and the moss *Funaria hygrometrica* with reference to the colonization of Surtsey. *Acta Bot. Is.* **4,** 10–15.

SPIESS L.C., LIPPINCOTT B.B. & LIPPINCOTT J.A. (1971) Development and gametophore initiation in the moss *Pylaisiella selwynii* as influenced by *Agrobacterium tumefaciens. Am. J. Bot.* **58,** 726–31.

SPIESS L.D., LIPPINCOTT B.B. & LIPPINCOTT J.A. (1976) The requirement of physical contact for moss gametophore induction by *Agrobacterium tumefaciens. Am. J. Bot.* **63,** 324–8.

SPIESS L.D. & LIPPINCOTT B.B. (1978) Association of moss and bacteria in nature. *Abstracts Bot. Soc. Am. Misc. Ser.* **156,** 91.

STEWART W.D.P. (1966) *Nitrogen Fixation in Plants*, p. 68. Athlone Press, London.

STEWART W.D.P. (1967) Transfer of biologically fixed nitrogen in a sand dune slack region. *Nature (Lond.)*, **214,** 603–4.

WEST G.S. (1908) Some critical green algae. *J. Linn. Soc. Lond. Bot.* **38,** 279–89.

WHATLEY M.H. & SPIESS L.D. (1977) Role of bacterial lipopolysaccharide in attachment to moss. *Plant Physiol. (Bethesda)*, **60,** 765–6.

CHAPTER 9
ECOLOGY

Various chapters of this book have described the ways in which mosses function, their response to environmental factors and their interrelationships with other organisms. It will already be realized that mosses are an extremely important part of the ecosystem in subarctic regions, in temperate and tropical forests and on mires. However, other habitats are also colonized and a mention of some of these is justified to show the interesting roles that mosses play in diverse situations. Explanation of the combination of environmental factors which enable a particular moss species or community to thrive in a certain niche (the essence of ecology) is beyond the scope of this text on mosses.

Suffice it to say that the factors include a combination of physical (light, temperature, wind, water etc.), edaphic (nature and type of substrate) and biotic (herbivores, parasites) components (Ando 1979; During 1979).

1 HOT SPRINGS

Iceland is famous for its volcanoes (see also Chapter 8, Section 4.1(a)) and hot springs. Water frequently emanates from the latter at close to 100°C but falls, within 100 m or so, to about half this temperature. The clouds of steam produced from the hot water in the cold environment ensure a constantly high humidity around the springs. Certain mosses are able to thrive alongside the water even close to the source and a zonation of species is observed. *Archidium alternifolium* (and some liverworts) form an inner zone, then tussocks of *Sphagnum* occur and finally a carpet of *Hypnum* spp. (Fig. 9.1; Richards 1932). The water temperature in the *Sphagnum* cushions may be as high as 42°C (Lange 1973). On geothermal sites in New Zealand, *Campylopus holomitrium* can colonize surfaces of mud which are at 50°C (Given 1980).

The occurrence of *Sphagnum* around hot springs is remarkable since the pH of the water is 7·5–9·0 and this moss is generally considered characteristic of acid water or soil (Fig. 9.2). However, the springs have

Fig. 9.1 The zonation of bryophytes by the outflow of a hot spring in Iceland. Nearest the water is a dark belt of liverworts and *Archidium*, then a broad belt of *Sphagnum* cushions and, outside that are *Hypnum* carpets (Richards 1932).

a very low calcium content but high levels of potassium, sodium and nitrogen gas, which suggests that *Sphagnum* may be more an avoider of calcium than a lover of acid conditions. In addition, H^+ ions released by the *Sphagnum* shoots may result in water within the *Sphagnum* tufts having a pH 3–4 log units lower than that of the surrounding water, as has been found in calcarious fens. Fifteen different species of *Sphagnum* grow around the springs and many which occur in this habitat are rare, or absent, elsewhere in Iceland. Such species are typical of warm oceanic climates and the hot springs seem to provide the necessary high humidity and warm temperature they require (Lange 1973).

A few of the hot springs gush forth acid water which is rich in sulphur and sulphur deposits occur close to the springs. A much poorer

Fig. 9.2 *Sphagnum tenellum* with abundant capsules. This species is typical of damp heaths and near the tops of hummocks in strongly undulating boggy ground (photo: M.C.F. Proctor).

bryophyte flora is found bordering hot acid water though *Sphagnum compactum*, a species which seems unusually tolerant of sulphur, does occur at a distance from such springs where the influence of the sulphur is reduced (Lange 1973; see also Chapter 10).

2 LAKES

At the bottom of lakes, mosses are often found growing on soft mud, sand and rock. The depths at which they occur depends on the penetration of light and this is affected by the amount of suspended particles, the concentration of dissolved humic acids (in water derived from peat deposits) or the thickness of ice cover, etc.

Lake Taho on the border of California and Nevada, USA, is one of the world's deepest lakes and has extremely clear water. In it, mosses have been found down to 150 m (Crum 1973). Another lake in America, Crater Lake, Oregon, has particularly dense growths of *Fontinalis antipyretica* and *Drepanocladus fluitans* in a sublittoral zone where the water depth is between 20 and 60 m. In arctic regions, deep water byrophytes have also been recorded down to about 30 m even in lakes which are permanently ice covered (Light & Lewis Smith 1976).

The Scottish lochs have been studied in some detail; those above 700 m elevation have clear rather than peat coloured water and may be covered with ice for 5 to 7 months of the year. Eight different species of moss have been collected at depths ranging from 1 to 20 m and some of these form dense luxuriant growths many square metres in extent. Several of the mosses are not normally regarded as aquatic, e.g. *Polytrichum commune* and *Sphagnum subsecundum* which usually grow on mires. Both may be found at depths of more than 5 m in a healthy condition and the former has been estimated to grow some 5 cm/yr in such situations. *Sphagnum subsecundum* from lake bottoms shows considerable morphological modifications from plants found on bogs so that at least one deep water form appears to have evolved in these Scottish lakes. Reproduction in deep water mosses is limited to vegetative propagation and although many almost certainly originated from propagules blown or washed in from terrestrial habitats, they have adapted well to the relatively stable deep water environment (Light & Lewis Smith 1976).

3 ROCK BUILDING

Rainwater slowly erodes limestone, yielding soluble calcium bicarbonate which is carried away by water percolating through the rock. If the water comes to the surface again via springs or rivers, re-deposition of the calcium carbonate may take place. A community of organisms, in which mosses play a key role, is responsible for re-deposition forming a rather soft material called tufa. With time, this hardens into a brownish limestone known as travertine (derived from the name of an ancient Roman town now known as Tivoli). This stone was used to build the Coliseum at Rome and often decorates the halls of the more elegant banks in America (Crum 1973).

In the formation of travertine, a thick spongy growth of mosses and algae develops on rocks in the rivers or around springs. These organisms provide surfaces which can absorb, retain, and expose copious thin films of calcium bicarbonate-containing water for effective evaporation, with the consequent loss of carbon dioxide and precipitation of calcium carbonate. The latter is in the form of calcite which hardens around the mosses taking a mould of their form. The lower portions of the moss cushion are progressively cemented while growth at the apex of the moss shoots proceeds resulting in a steady increase in the thickness of the tufa (Parihar & Pant 1975).

Deposits of travertine more than 30 m thick are known but perhaps the most remarkable occur in Burma not far from Mandalay. There, the deposits take the form of weirs from a few centimetres to more than 2 m in height and are as level along the crest as if built by human hand (Fig. 9.3). The dams occur at points along the rivers where one would expect

Fig. 9.3 A series of tufa (travertine) dams in Ke-Laung Stream, Wetwin, Burma whose development is dependent on the growth of bryophytes (La Touche 1913).

to find rapids. On the weirs, growth of the tufa is most rapid along the lip where the water is most agitated, presumably because the moss, which is found wherever tufa is being deposited, is able to grow most rapidly there (La Touche 1913). In this area of Burma, travertine deposits are also found beside every spring so that the scarped cliffs along faults and precipitous gorges are festooned with moss covered curtain-like masses. These may extend across a gorge and form natural bridges up to 170 m above the river as at Gokteik midway between Maymyo and

Lashio. This bridge was a traditional crossing point for the Chinese transporting goods by mule but was subsequently used as the support for a railway viaduct whose height where it rests on the natural bridge is 100 m (La Touche 1906).

In the northern temperate zones, the chief tufa-forming mosses are *Barbula tophacea*, *Cratoneuron* spp., *Eucladium verticillatum*, *Gymnostomun* spp., and *Philonotis* spp. while in India and Burma, *Barbula*, *Bryum* and *Hymenostiella* are important (Emig 1918; Parihar & Pant 1975). In India, the filamentous blue-green alga *Petalonema* is epiphytic on the mosses, while other unicellular forms, the green alga *Cladophora*, diatoms, liverworts, and a few flowering plants complete the tufa-forming community. The latter help in building the tufa, but only do so when their exposed stems are covered with a felt of bryophytes and algae. Another moss, *Brachythecium rivulare* is associated with a rather unusual form of highly iron-impregnated tufa around mineral springs in Indiana and Illinois, U.S.A. As iron compounds penetrate the moss tissues, a hard, porous tufa is formed, which becomes part of the accumulation of bog iron ore around the springs. Together with other species of plants, the moss may build up such high mounds around the spring that it is necessary for the water to find a new exit point (Taylor 1919).

4 SEA WATER

In general, mosses avoid seawater and none are marine. However, there are a few mosses which are obligate halophiles growing an coastal rocks just above high tide marks where they may be periodically inundated during storms and regularly covered with spray. *Dicranella siliquosa*, a halophile from Japan, may, even when washed, contain 28% sodium chloride (Richards 1932). Recent studies in England on the seashore mosses *Grimmia maritima*, *Pottia heimii*, *Tortella flavovirens* and *Ulota phyllantha* have shown that they can withstand more than 24 h immersion in seawater without harmful effects such as the loss of intracellular potassium or reduction in rates of photosynthesis. Calcium and to a lesser extent magnesium ions are involved in the observed tolerance to seawater since loss of potassium is observed in equivalent concentrations of pure sodium chloride. Inland mosses, unlike those from the seashore, are unable to maintain their intracellular potassium levels when immersed in seawater and show other symptoms of damage. *Ulota phyllantha* is a supralittoral species which is rather less tolerant than the other seashore mosses. Populations of *Ulota phyllantha* from the seashore however, have a greater resistance to the effects of seawater than those growing on trees a short distance inland. Species like *Grimmia maritima* maintain levels of essential intracellular metal ions, such as potassium, by having a low overall permeability to metal ions, so the high internal levels do not build up quickly on exposure to seawater, and by possessing a sodium pump which extrudes sodium from the moss cells (Bates & Brown 1974, 1975; Bates 1976).

Salt marshes are another habitat where plants have to withstand exposure to seawater. More than 50 moss species were found in a survey of the vegetation of British salt marshes. Few form sporophytes except *Pottia heimii* which is so prolific in this regard that the moss can be distinguished at a distance by the brown patches on the marsh due to the innumerable capsules. Salt marsh mosses which grow in situations covered more than 100 times a year by the tide, include *Pottia heimii*, *Amblystegium serpens* and *Eurhynchium praelongum* in England. *Campylium polygamum* and *Grimmia maritima* also occur in this zone in Scotland (Adam 1976). Certain mosses are very common on a few salt marshes, where they are abundant, but are absent from many others, *Archidium alternifolium* being a striking example. It will be recalled that this moss was the one that occurred closest to the alkaline hot spring water in Iceland, but the conditions which promote its occurrence in two such different habitats are unknown (Adam 1976).

A few mosses grow where the salty water in their environment is derived from minerals. For example, *Funaria hungaria* lives on the salt lands in Eastern Europe, *Pottia crinita* is found near salt deposits in Germany and *Funaria sickenbergii* grows among crystals of magnesium sulphate on the edge of the Egyptian shots (Richards 1932). Mosses are also found in hollows associated with salt extraction in Cheshire, England, which have become contaminated by residues from the ammonia–soda process of alkali manufacture. *Barbula tophacea* (usually a tufa-forming moss) and *Funaria hygrometrica* (normally a moss of burnt over ground) are found where concentrations of magnesium exceed those of calcium and the pH is 8·4–9·3. *Pohlia annotina* (a common moss of salt marshes) occurs where the levels of calcium are greater than magnesium, the pH 7·9–8·8 and the through-flow of freshwater greater (Mizra & Shimwell 1977; Joenje & During, 1977).

5 SAND

5.1 COASTAL DUNES

Extensive sand dune systems occur along many sea coasts of the world and also on the margin of some of the Great Lakes in North America particularly Superior, Michigan, and Huron. Sand dunes may be more than 30 m high and many square km in extent, with marram grass usually being of key importance in stabilization of the sand in temperate zones. A succession of plants occurs on dune systems, frequently culminating in deciduous or coniferous forest. Mosses play a very important role in dune colonization, helping to stabilize the sand, adding organic matter and maintaining moist conditions near the sand surface through the water holding capacity of the gametophores and rhizoid wefts.

The mobile 'yellow' dunes close to the water are colonized by a few

Table 9.1 Common sand dune mosses in Scotland

Species	Habitat
Bryum pendulum *Tortula ruraliformis* *Ceratodon purpureus*	Species of mobile dunes
Brachythecium albicans	Rapidly becomes prominent amongst the preceding species
Pseudoscleropodium purum *Rhytidiadelphus squarrosus* *Climacium dendroides* *Hylocomium splendens* *Rhytidiadelphus triquetrus* *Hypnum cupressiforme* *Dicranum scoparium* *Pleurozium schreberi* *Polytrichum juniperinum* *P. piliferum* *Pohlia annotina* *Hypnum cupressiforme* var. *tectorum*	Species of fixed dune and dune pastures Species of dune heath
Bryum argenteum	Grey dunes and dune slacks
Rhacomitrium canescens	Dry hollows and blow-outs

Adapted from Birse *et al.* (1957)

species of mosses (Table 9.1) which grow between the shoots of the dune-building grasses. Such mosses have the ability, not only to grow through up to 4 cm of sand deposited on them, but also quickly to re-establish a moss colony on emerging. For example, the shoots of *Ceratodon purpureus* grow through the sand, at the same time producing upwardly extending rhizoids which form protonemata on reaching the surface. Buds on the latter develop a dense stand of shoots and a compact turf is thereby formed (Birse *et al.* 1956).

Tortula ruraliformis is the moss that grows fastest through a one cm covering of sand and possesses other strategies of advantage to a dune moss. The leaves which have hair points wrap themselves around the stem in dry weather in a tight spiral. When the dunes are dry, sand is blown about and the plants are likely to be covered. As soon as rain moistens the sand, the plants absorb water and the leaves untwist so violently that sand grains are thrown to a distance of several centimetres. In this way each shoot cleaves a little pit in the overlying sand. *Tortula ruraliformis* occurs on mobile dunes in carpets of a characteristic shape: a large round central patch with single plants or smaller patches some centimetres away. Rhizoids run out from the central plant terminating in protonemata and buds that develop into the satellite clumps. In this way, the moss can rapidly extend over the surface of a dune (Richards 1929). *Brachythecium albicans* behaves in yet a different manner in that shoots emerging from a sand covering turn through 90°, growing prostrate along the sand. They form abundant rhizoids and branches so that a fairly compact mat quickly develops (Birse *et al.* 1956).

The fixed dunes are colonized by mosses which come in as soon as the rate of sand deposition is curtailed and the surface stabilized. Such mosses are larger, more robust, and less firmly anchored to the substratum (Table 9.1). Experimental studies have shown that they are often able to survive being covered by sand, but grow slowly on emergence. Undefined environmental or competition factors result in the fixed dune species being replaced by carpets of dune heath mosses which, in the case of *Hypnum*, *Rhytidiadelphus* and *Dicranum*, show negligible ability to penetrate a sand covering. However, it is interesting that *Pohlia annotina* and *Polytrichum juniperinum*, typical of dune heaths, can emerge from deep burial but are distinctly slower to re-establish themselves than other sand dune mosses. This may be one factor accounting for their absence earlier in the dune succession (Birse *et al.* 1956). *Polytrichum* is also far less tolerant of salt spray than *Ceratodon purpureus* which grows on the mobile dunes close to the sea (Boerner & Forman 1975).

It should be noted that while mosses are pioneers among the lower plants in dune systems, they are often replaced by lichens as ground cover at an intermediate stage. Lichen genera such as *Peltigera, Cornicularia, Cladonia* and *Cetraria* are particularly common, but even on dunes where they are abundant, mosses return later in the succession as they seem better able to compete with higher plants than lichens (Richards 1929).

5.2 Hot deserts

The occurrence of mosses in hot deserts is remarkable when one considers that they require water for sperm transfer and hence completion of their life cycle. Mosses cannot draw water from underground reserves, or even retain it for ver very long. Their ability to absorb water from dew or humid night air allows them to build up a sufficient water content for a brief period of photosynthesis in desert habitats following dawn. Like the lichens, they dry out as the morning progresses and stay desiccated until sunset, when the cycle begins once more (Kappen *et al.* 1979).

In the hot deserts of North America both the mosses and lichens are virtually restricted to north-facing slopes where surface temperatures are in approximate equilibrium with the air temperature. South-facing slopes are too arid for these plants, because the surface temperature can increase by as much as 20°C within the first 30 min after sunrise causing any moss there to dry out extremely rapidly. The abundance of mosses in these deserts is consistently less than lichens (maximum of 0·3% v. 8% cover) with a biomass of only about 2 g/m^2. At least 18 different species of moss grow in the hot deserts of North America and most, such as *Tortula intermedia, Grimmia laevigata* and *Bryum caespiticium* are tuft forming (Nash *et al.* 1977).

In the Negev desert, Israel, there are rather more mosses, since higher humidities and more frequent dew formation occur due to the proximity of the Mediterranean and Red Seas. In all deserts, disturbance by animals or man can seriously reduce the amount of moss and lichen growth by destroying the fragile surface crusts on which many species grow.

6 FIRE

Every year large areas of grassland and temperate and tropical forest catch fire, some through accident or natural causes, others by man's agency as part of agricultural or forestry practice. The resulting tracts of land provide habitats for a succession of mosses but they only become abundant if colonization by higher plants is slow, as on heaths, in upland areas, or at higher latitudes.

A key factor affecting colonization of burned sites by mosses is the rapidity of the fire. Conflagrations which spread rapidly, e.g. over grassland, deposit little ash and do not greatly change the soil conditions as the soil surface is exposed to high temperatures for only a short duration. The mosses and flowering plants which subsequently colonize tend to be those that were present in the pre-burn vegetation. In contrast, slow fires deposit large amounts of ash. They raise the surface temperature up to 900°C, the layer below to about half this and greatly modify the soil. Thus the pH may rise from 5–6 to 9–11 with high levels of potassium, magnesium, calcium, phosphate, various forms of nitrogen and soluble organic matter. On sites subject to this sort of fire, a specialized and abundant moss flora usually develops (Southorn 1976).

In England, on experimentally produced bonfire sites having a covering of 2–3 cm of ash, moss protonemata appear in the soil as little as 9 weeks after the fire if it occurs in early spring; gametophores

Fig. 9.4 Recently felled woodland. The area where the brushwood was burned is covered by a carpet of the moss *Funaria hygrometrica* (photo: J.G. Duckett).

develop in 4–5 months (Fig. 9.4). If the fire takes place in autumn colonization is slower. *Funaria hygrometrica*, *Ceratodon purpureus*, *Bryum argenteum* and other *Bryum* species with rhizoid gemmae (tubers) form the pioneer moss community. However, *Funaria* rapidly becomes dominant, the other species remaining as scattered shoots among this moss (Fig. 9.5). Competition from flowering plants, which tend to spread from the margin of the burned area, leads to the disappearance of the mosses but if the sites are weeded, the pioneer mosses continue to thrive. If ash is removed from a bonfire site and placed on an area weeded free of higher plants, then a typical bonfire moss community develops. In contrast, on bonfire sites from which all ash is

Fig. 9.5 A close up photograph of the capsules of *Funaria hygrometrica* (photo: M.C.F. Proctor).

immediately removed, *Bryum argenteum* becomes most abundant followed by *Ceratodon purpureus*, but *Funaria hygrometrica* is never dominant. This suggests that only *Funaria hygrometrica* has a distinct preference for burned sites (indeed dense tufts of this moss are seldom found elsewhere). The other species have a wide ecological amplitude and simply take advantage of the open space provided by the fire (Southorn 1976). These mosses often occur in great abundance in unburned but disturbed substrata. *Bryum* spp. in particular seem to favour habitats enriched with soluble organic matter, as shown by their abundance on moist lawns when the grass cuttings are left to decompose, and even on the rim of the trickle filters used in sewage treatment plants (Bridge Cooke 1953). These mosses also grow on cooling towers of industrial plants where they corrode the concrete and can utilize nutrients in the water including polyphosphates, acetone and phenols (Munjko *et al.* 1973).

In North America and elsewhere, *Polytrichum* sp. often succeeds the pioneer moss community, as the nutrient status of the burnt soil decreases due to leaching of the soluble ions (K^+ ions fall rapidly over a period of weeks, then Mg^{2+}, while Ca^{2+} scarcely changes over 18 months) (Southorn 1976). At one site in Michigan, *Ceratodon purpureus* was first observed in 1930 four years after the fire. By 1940, this moss covered 50% of the ground, and by 1950, it had colonized 95%. *Polytrichum piliferum* appeared in 1942 and by 1971, in association with *Cladonia* lichens, had almost replaced *Ceratodon purpureus*.

Another moss which is a transient of burned sites in both North America and Europe is *Buxbaumia aphylla*. This is usually found 15–50 years after a fire, in areas where the heat was extremely intense, and at the successional stage at which *Cladonia* lichens are common. *Buxbaumia aphylla* has also been found to be a frequent colonizer of colliery debris in Lanarkshire, Scotland (Corner 1968). Although pit heaps sometimes smoulder if they contain much coal, there would appear to be little else in common between mine debris and forest or heath fire sites.

7 TREES

A tree is a complex habitat with the character of the bark changing from the youngest twigs to the old trunk. The attachment and establishment of bryophtes are affected by this feature. The light intensity falling on the various surfaces of a tree also varies, being greatest in the crown and on the south side of the trunk in the northern hemisphere. However, the geometry of a tree's crown and the sun's pathway may combine to give a complex distribution of light and shade on the trunk. An important factor controlling the amount of epiphytic growth is the water supply which is intermittent. The mosses growing on the upper sides of sloping branches facing the prevailing wind receive the greatest amount of rain and run-off while overhung or leeward surfaces of the tree are driest. Four broad habitats for epiphytes can frequently be distinguished on trees: the base of the trunk in close proximity to the soil, the trunk, the larger spreading branches and the younger twigs (Proctor 1962).

7.1 Tropical forests

The communities of mosses on trees in these forests are complex and little studied but some moss species show a particular preference for one kind of tree. Thus, in the rain forest of Guiana, *Moenkemeyera richardsonii* is usually found on *Miconia plukenetii* whose bark has a characteristic texture (Richards 1932). Perhaps the most remarkable mosses of tropical trees are those which grow on long-lived leaves. Few moss species are truly epiphyllous but often grow onto the leaves from the neighbouring twigs. Liverworts and lichens are also common and all prefer tough leathery leaves, avoiding those with non-wettable surfaces. Usually the young leaves are first colonized by lichens, followed by liverworts and mosses. Apparently epiphyllous liverworts, such as *Radula flaccida* which grow on tree leaves in West Africa, are able to absorb water and phosphate from the host leaves. However even the heaviest colonization only reduces the light reaching the leaf by some 2% and photosynthate does not pass from host to epiphyte or vice versa. Thus little harm is likely to accrue from such growths (Eze & Berrie 1977). The oldest leaves may have such a dense felt of these plants that seeds of epiphytic flowering plants germinate among them (Richards 1932). Monocotyledons as well as dicotyledons may be colonized. In Malaya, the moss *Ephemeropsis tjibodensis* grows on the leaves of monocotyledons forming horizontal protonemata with assimilatory side branches that grow upwards and terminate in long bristles. In Queensland, Australia this moss grows on lawyer vines and other broad leaved trees (Stone pers. comm.). Holdfasts from the basal protonemata attach the plant to the leaf surface. Not surprisingly, the gametophores are very reduced and the capsules quite small (Bower 1935).

7.2 Temperate forests

Mosses are most luxuriant in forests of moist climates and several communities may be recognized growing epiphytically on trees in such situations (Barkman 1958). For example, three clear bryophyte zones may be found on the oak trees in the high altitude (350–450 m) woodland on Dartmoor in western Britain. Here rainfall is about 1,500 mm per year.

The trunks and lower branches of the oaks are covered by an abundant growth of bryophytes particularly the mosses *Dicranum scoparium* and *Isothecium myosuroides*. The former is the pioneer species and *Isothecium* then colonizes to build up a dense mat which later is invaded by liverworts including the conspicuous brown species *Frullania tamarisci*. Eventually this mat with its considerable accumulation of humus and windblown soil particles begins to break up. Bryophytes characteristic of peat and decaying wood, including *Tetraphis pellucida*, then appear. Finally the bare trunk is re-exposed and colonization by the pioneer species *Dicranum scoparium* begins once more (Proctor 1962).

The higher branches in these Dartmoor woodlands become covered with *Hypnum cupressiforme* but scattered colonies of *Dicranum scoparium* are found together with foliose and fruticose lichens.

The uppermost twigs are colonized by tuft-forming mosses such as *Ulota bruchii* and *Ulota crispa* with sparse amounts of the liverwort *Frullania dilatata*. Epiphytes first appear on the twigs when they are 8–9 years old. *Dicranum scoparium* begins to colonize when the twigs reach 4–6 cm in diameter (15–18 years old). Slender branches dominated by *Ulota* and *Frullania* are sometimes seen well below the canopy. These grow out of the larger branches which are thickly covered with *Hypnum*. This shows that the borderline between the different zones is not simply a question of light intensity (see Chapter 3, Section 1) but is determined partly by light, partly by water supply and partly by bark character (Proctor 1962). The distribution patterns of bryophytes is further complicated by historical considerations such as the impact of the early quaternary climate, the rate of spread following the retreat of the ice after the last glaciation and recent fragmentation of the range of different species via human interferences such as forest clearance (Ratcliffe 1968; Dickson 1973).

8 REFERENCES

ADAM P. (1976) The occurrence of bryophytes on British saltmarshes. *J. Bryol.* **9,** 265–74.

ANDO H. (1979) Ecology of terrestrial plants in the Antarctic with particular reference to bryophytes. *Mem. National Instit. Polar Res. Tokyo, Special Issue,* **11,** 81–103.

BATES J.W. & BROWN D.H. (1974) The control of cation levels in seashore and inland mosses. *New Phytol.* **73,** 483–95.

BATES J.W. & BROWN D.H. (1975) The effect of seawater on the metabolism of some seashore and inland mosses. *Oecologia (Berl.)*, **21,** 335–44.

BARKMAN J.J. (1958) *Phytosociology and Ecology of Cryptogamic Epiphytes.* Assen, Netherlands.

BATES J.W. (1976) Cell permeability and regulation of intracellular sodium concentration in a halophytic and glycophytic moss. *New Phytol.* **77,** 15–23.

BIRSE E.M., LANDSBERG S.Y. & GIMININGHAM C.H. (1957) The effects of burial by sand on dune mosses. *Trans. Br. Bryol. Soc.* **3,** 285–301.

BOERNER R.E. & FORMAN R.T.T. (1975) Salt spray and coastal dune mosses. *Bryologist*, **78,** 57–63.

BOWER F.O. (1935) *Primitive Land Plants,* pp. 74–7. Macmillan, London.

BRIDGE COOKE W. (1953) Mosses in a sewage treatment plant. *Bryologist*, **56,** 143–5.

CORNER R.W.M. (1968) Further observations on the distribution of *Buxbaumia aphylla* Hedw. on colliery debris. *Trans. Br. Bryol. Soc.* **5,** 828–9.

CRUM (1973) Mosses of the Great Lakes Forest Region. *Contrib. Univ. Mich. Herb.* **10,** 1–404.

DICKSON J.H. (1973) *Bryophytes of the Pleistocene.* Cambridge University Press.

DURING H.J. (1979) Life strategies of bryophytes: a preliminary review. *Lindbergia*, **5,** 2–18.

EMIG W.H. (1918) Mosses as rock builders. *Bryologist*, **21,** 25–7.

EZE J.M.O. & BERRIE G.K. (1977) Further investigations into physiological relationships between an epiphyllous liverwort and its host leaves. *Ann. Bot. (Lond.)*, **41,** 351–8.

GIVEN D.R. (1980) Vegetation on heated soils at karapiti, central North Island, New Zealand, and its relation to ground temperature. *N.Z. J. Bot.* **18,** 1–13.

JOENTE W. & DURING H.J. (1977) Colonisation of a desalinating wadden-polder by bryophytes. *Vegetatio*, **35,** 177–185.

KAPPEN L., LANGE O.L., SCHULZE E.D., EVENARI M. & BUSCHBOM U. (1979) Ecophysiological investigations on lichens of the Negev desert. 6. Annual course of photosynthetic production of *Ramalina maciformis* (Del.) Bory. *Flora (Jena)*, **168,** 85–108.

LANGE B. (1973) The *Sphagnum* flora of hot springs in Iceland. *Lindbergia* **2,** 81–93.

LATOUCHE T.D. (1906) Note on the natural bridge in Gotkeik Gorge *Rec. Geol. Surv. India*, **33,** 49–54.

LATOUCHE T.D. (1913) Geology of the Northern Shan States. *Mem. Geol. Survey India*, **39,** 325–9.

LIGHT J.J. & LEWIS SMITH R.I. (1976) Deep-water bryophytes from the highest Scottish lochs. *J. Bryol.* **9,** 55–62.

MIZRA R.A. & SHIMWELL D.W. (1977) Preliminary investigation into the colonization of alkaline industrial waste by bryophytes. *J. Bryol.* **9,** 565–72.

MUNJKO L., TOMEC M. & PAVLETIĆ Z. (1973) Neke ekološke karakteristike mahovina rashladnog tornja Organsko kemijske industrije u Zagrebu. (Some ecological charactersistics of the mosses from the cooling tower of petrochemical works in Zagreb.) *Acta. Bot. Croat.* **32,** 117–23.

NASH T.H., WHITE S.L. & MARSH J.E. (1977) Lichen and moss distribution and biomass in hot desert ecosystems. *Bryologist*, **80,** 470–9.

PARIHAR N.S. & PANT G.B. (1975) Bryophytes as rock builders— some calcicole mosses and liverworts associated with travertine formation at Sahasradhara, Dehra Dun. *Current Science*, **44,** 61–2.

PROCTOR M.C.F. (1962) The epiphytic bryopyhte communities of the Dartmoor oak woods. *Trans. Devonshire Assoc.* **94,** 531–54.

RATCLIFFE D.A. (1968) An ecological account of atlantic bryophytes in the British Isles. *New Phytol.* **67,** 365–439.

RICHARDS P.W. (1929) Notes on the ecology of bryophytes and lichens at Blakeney Point, Norfolk. *J. Ecol.* **17,** 127–40.

RICHARDS P.W. (1932) Ecology. In *Manual of Bryology*, ed. Verdoorn F. pp. 367–95. Martinus Nijhoff, The Hague, Holland.

SOUTHORN A.L.D. (1976) Bryophyte recolonization of burnt ground with particular reference to *Funaria hygrometrica.* I. Factors affecting the patterns of recolonization. *J. Bryol.* **9,** 63–80.

TAYLOR A. (1919) Mosses as formers of tufa and of floating islands. *Bryologist,* **22,** 38–9.

CHAPTER 10
AIR POLLUTION

The most important air pollutants in terms of bryophytes are sulphur dioxide, lead, hydrocarbons and nitrogen oxides. Photo-oxidation of the latter two may give rise to harmful concentrations of ozone. The major gaseous pollutants from industrial sources are sulphur dioxide (SO_2), from power stations and smelters, and gaseous fluorides (HF and SiF_4) from aluminium smelters and brick works. In addition, metallic pollutants, frequently in the form of metal oxides, sulphates or sulphides, are often released in large quantities from industrial processes.

In general, SO_2 causes the most widespread damage to mosses; the effects of gaseous fluorides are more localized and very little is known about the impact of ozone on mosses under natural conditions. Metals may exacerbate the effects of gaseous pollutants emitted simultaneously or may themselves induce abnormal symptoms close to pollution sources. The overall result is that the moss flora of an area subject to air pollution is much impoverished. For example, 23 bryophyte species which occurred in Amsterdam in 1900 are now extinct there, and comparable changes have been observed in England due to man's activities (Barkman 1969; Rose & Wallace 1974).

1 GASEOUS POLLUTANTS

1.1 SO_2 Uptake mechanisms

Only SO_2 will be considered because more is known about the uptake of this gas and it is the principal one which affects mosses. Poikilohydrous plants such as mosses and lichens are insensitive to SO_2 when dry, and this is probably the main reason why the distribution of sensitive species correlates best with the mean winter SO_2 concentrations in temperate climates. The considerable SO_2 sensitivity of mosses, especially in winter, is related to three aspects of their biology. Firstly, they are metabolically most active during the cool moist periods of the year and hence SO_2 exposure at this time is likely to interfere significantly with growth and reproduction. Secondly, moss gametophores do not possess stomata or thick waxy cuticles which help to exclude pollutant gases from flowering plants at least during the hours of darkness. Thirdly, many mosses are moist, or may, as in the case of *Sphagnum*, be covered with a thin film of water when they are physiologically active. Sulphur dioxide is a very soluble gas and will continuously dissolve in the moisture until an equilibrium is reached according to the equation:

$$\text{ppm } SO_2 \text{ (wt/wt), in water} = 10.3\sqrt{SO_2} \text{ (vol./vol.) ppm, in air}$$

For example, an atmospheric concentration of 0·5 ppm (1300 $\mu g/m^3$) could give rise to a solution containing 7 ppm of dissolved SO_2 in the thin water film covering the plant (Nieboer *et al.* 1977; Tomassini *et al.* 1977). Thus the poikilohydrous nature of mosses (and lichens) facilitates the accumulation of high levels of dissolved SO_2 and so it is not surprising that they are among the most SO_2 sensitive plants. The dissolved SO_2 may affect the cell membranes resulting in leakage of cytoplasmic constituents such as K^+, or it may be taken up by the cells of the moss.

Dissolved SO_2 forms the anions HSO_3 and SO_3^{2-} (see below) and these, like other anions, are taken into cells via a process which involves active transport (i. e. expenditure of cell energy), and is mediated by a carrier with the properties of an enzyme. Transport is unidirectional from outside to inside the cell, and the rate of transport is governed by pH, temperature, concentration of the anion, and the availability of metabolic energy within the cell (Schiff & Hodson 1973; Ziegler 1975). In contrast to the uptake of sulphate (SO_4^{2-}), sulphite (SO_3^{2-}) becomes covalently bound within the cell to sulphydryl groups directly or indirectly associated with photosynthetic electron transport. There appears to be no effective control step which limits the uptake of SO_3^{2-} comparable to that which operates for sulphate. Thus sulphite readily accumulates within cells (Ziegler & Hampp 1977) and this can lead to deleterious consequences such as a reduction in photosynthesis (Nieboer *et al.* 1976).

1.2 EFFECTS OF SULPHUR DIOXIDE

In temperate climates such as Western Europe, most bryophytes do not exist where the winter mean concentration of SO_2 exceeds 50$\mu g/m^3$ (0·17 ppm). Some mosses, e.g. *Ceratodon purpureus* and *Bryum argenteum* seem able to tolerate much higher pollution levels than others, e.g. *Grimmia pulvinata* and *Hypnum cupressiforme* (Gilbert 1968). Sensitive species, transplanted into an industrial area containing toxic amounts of

SO_2, show symptoms of damage which begin with the loss of green colour at the tips of the more exposed leaves. This colour change gradually extends down the shoots and, completely bleached specimens usually fail to recover even if returned to a pollution free site, although regeneration from the basal portions of the moss may occasionally occur. These acute symptoms result from degradation of chlorophyll and damage to cell membranes which leads to plasmolysis of the cells (LeBlanc & Rao 1974).

Low levels of SO_2, while not causing the acute symptoms described above, can still affect the diversity and abundance of naturally growing mosses. The disappearance of the once dominant *Sphagnum* species from upland blanket bogs (mires) of the Southern Pennines in England is thought to be due to the effects of sulphur dioxide dissolved in the rain (acid rain). *Sphagnum recurvum* is the one species which exists today in considerable quantities in that area and its presence may be related to the ability it exhibits in laboratory experiments to withstand the effects of relatively high concentrations of SO_2 (Ferguson *et al.* 1978).

In Canada, low levels of SO_2 derived from natural gas purification plants were found to decrease the thickness and biomass of moss cushions growing beneath white spruce trees. More severe stress close to the SO_2 source is characterized by a reduction in percentage cover and capsule production by the moss as compared with control areas (Winner & Bewley 1978). Concentrations as low as 0·068 ppm in outdoor perspex chambers experimentally fumigated for 26 weeks also resulted in a drop from 70 to 20% in the cover of *Bryum* spp. growing below ryegrass (Bell 1973). The reduction in biomass and cover may result from effects on moss protonemata rather than on the gametophores. It has been shown in the laboratory that protonemata are some ten times more sensitive to SO_2 (Gilbert 1968; Nash & Nash 1974; Ferguson & Lee 1979). Indeed SO_2-resistant species of moss show the common features of fast growth and short-lived protonemal stages that quickly initiate buds forming the more resistant gametophores (LeBlanc & Rao 1974). Mosses have been shown experimentally to be even more sensitive to SO_2 than lichens (Inglis & Hill 1974; Türk & Wirth 1975) though the latter are regarded, from field observations, to be amongst the most sensitive of plants (Hawksworth & Rose 1976).

The toxicity of SO_2 results from the fact that the gas forms highly reactive species on dissolving in water ('H_2SO_3', HSO_3^-, SO_3^{2-}). The most toxic forms seem to be 'H_2SO_3' and HSO_3^- which form when the pH of the solution is low. Thus 0·1 mM HSO_3^- reduces, almost instantaneously, the evolution of oxygen by photosynthesizing *Sphagnum* plants. This is not reversed by adding HCO_3^- (dissolved carbon dioxide in the form of bicarbonate ion) so that more than competitive inhibition of the sites on enzymes involved in carbon dioxide fixation, is involved. SO_3^{2-}, the ion formed by dissolved SO_2 in solutions buffered to high pH, has no effect on oxygen evolution by *Sphagnum* under comparable conditions. Thus the pH of the substrate, or environment in which a moss is growing, may be an important factor and indeed the pollution sensitive moss *Grimmia pulvinata* occurs on asbestos-cement roofs, whose pH is high, 6 km from the centre of the industrial city of Newcastle upon Tyne, England, but does not grow on sandstone wall tops within 11 km (Gilbert 1970). However, the calcium in the asbestos-cement may provide additional SO_2 protection (Nieboer *et al.* 1979; Richardson *et al.* 1979a). Finally, it is unlikely that the disappearance of mosses in the field is simply due to the effects of SO_2 on photosynthesis, because this gas when dissolved has the potential for interfering in a deleterious way with many aspects of a cell's metabolism (Nieboer *et al.* 1976).

1.3 Effects of fluorides and ozone

Little work has been done on the effects of these gases on mosses. Around an aluminium smelter at Fort William, Scotland which emits fluorine, *Pohlia nutans* and *Aulacomnium androgynum* together with most terricolous species seem comparatively resistant. Sensitive epiphytic mosses such as *Orthotrichum* spp. and *Ulota crispa* exhibit browning of the leaf tips and this was also found in other mosses growing very close (within 180 m) to the smelter (Gilbert 1971). Similar symptoms were also observed on the leaves of *Polytrichum commune* growing around a phosphorus plant at Long Harbour, Newfoundland, Canada. The phosphate rock used at the plant contains about 4% of fluorides and as a consequence some gaseous fluorides are released from the reduction process into the atmosphere. The degree of discolouration observed in the moss leaves was directly correlated with the fluoride concentration (up to 6,000 ppm dry wt) in the moss tissue and inversely correlated with distance of the collection site from the phosphorus plant. Five per cent of the moss leaves had discoloured tips when the fluoride level in them was about 65 ppm dry weight while over 90% of the leaves showed severe browning when the levels reached 4,500 ppm dry wt. A reduction in capsule production was also noted in response to increasing pollution levels near the phosphorus plant. It is interesting to note that the fluoride content of the moss was almost ten times that of balsam fir needles collected from the same site (Roberts *et al.* 1979).

The degree to which ozone affects mosses has not been examined in detail but this gas does reduce the ability of detached leaves of *Funaria hygrometrica* to regenerate protonema (Comeau & Leblanc 1971).

2 METALLIC POLLUTANTS

2.1 UPTAKE MECHANISMS

Mosses have evolved efficient uptake mechanisms for absorbing metals and other nutrients from their environment. This feature is of advantage to mosses, particularly epiphytes, growing in nutrient-poor habitats. Those mosses which are endohydric take up metal ions principally at the base of the shoot or cushion and transfer them to the growing apex. In contrast, ectohydric mosses take up ions over their entire surface. The greater part of the metal content of mosses is, however, not taken up and located within the cells but is accumulated extracellularly in two ways, i.e. via ion exchange and particulate trapping.

(a) *Ion exchange*

The cell walls of mosses possess ion exchange sites onto which metal ions bind in a way comparable to the binding of ions onto synthetic ion exchange resins. If a moss (or a lichen) is placed in a solution of a metal salt, ion exchange takes place in a matter of minutes and does not require the expenditure of metabolic energy. The following equilibrium describes ion exchange in which metal ions are being taken up and hydrogen ions released:

$$M^{2+} + 2HA = MA^{+} + A^{-} + 2H^{+}$$

where M^{2+} denotes the entering metal ion, H^{+} the hydrogen ion released, HA the fixed protonated anionic functional group (ion exchange site) and MA the metal ion bound to the functional group (Nieboer *et al.* 1978).

Sphagnum is a most remarkable moss because large numbers of protonated anionic functional groups (ion exchange sites) in the form of uronic acids are formed during the synthesis of new wall material at the growing plant apex. In *Sphagnum* growing in its natural habitat, metal ions replace the protons (as described in the above equation), with the result that hydrogen ions are released. These ensure the maintenance of the low pH that is ideal for the development of this moss and which helps to inhibit competing species (Clymo 1963). Twenty to thirty per cent of the dry weight of *Sphagnum* may be uronic acids, so that the moss has a large cation binding capacity (900–1500 μmol/g), which is about 1/10 to 1/3 that of synthetic ion exchange resins (Richardson *et al.* 1980).

True mosses also possess metal binding sites though they are less abundant than in *Sphagnum* so that the uptake capacity is smaller. In both groups of mosses, bound metals can be displaced experimentally with H^+ (by shaking the moss in a dilute acid solution), thus driving the equilibrium in the above equation from right to left. Alternatively, by shaking in a suitable concentration of a second metal, the originally bound metal can be displaced because of the greater concentration of the second or its greater affinity for the binding site (Nieboer *et al.* 1978; Brown & Buck 1978).

(b) *Particulate trapping*

The surface of moss gametophores, when viewed with the scanning electron microscope, is often rough and provides unnumerable crevices for the entrapment of tiny particles. *Dicranum scoparium*, whose surface is minutely hairy (tomentose), traps more radioactively tagged fly-ash particles than *Polytrichum commune*. This ability to trap particles is also exhibited by *Dicranum* sp. in the field. Around a uranium mining and milling complex at Elliot Lake Ontario, samples of this moss with a high titanium content also have a high iron content. This is not surprising if the moss is trapping tiny particles of rock in which the ratio of the two metals is constant, and indeed this occurs because oxides of iron and titanium tend to crystallize together in rocks. If the iron content of a series of moss samples is plotted against their titanium content, then a straight line relationship is observed (Fig. 10.1) (Richardson *et al.* 1979b). However a change in the Fe/Ti ratio would be expected and is observed if the particulates accumulated by a moss are enriched with one or other metal. Thus, near a steel refinery there will be a large proportion of particles from the industry which have a high iron content while around an oil fired power station there will be many particulates rich in titanium (Nieboer *et al.* 1978).

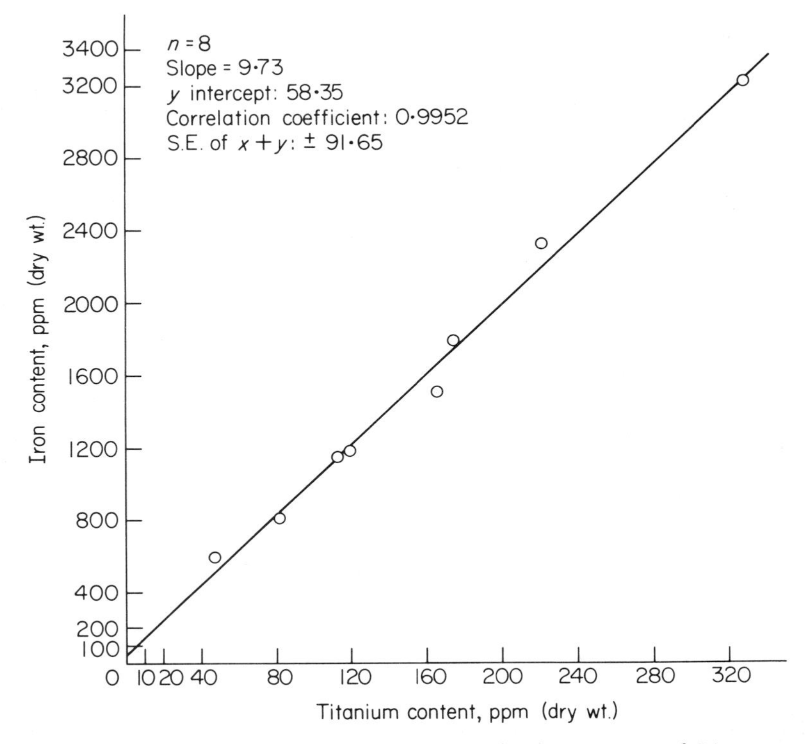

Fig. 10.1 The linear relationship between the iron and titanium contents of *Dicranum* spp. from around Elliot Lake, Ontario which indicates particulate trapping by mosses (Boileau, Richardson & Nieboer unpublished data).

2.2 Effects of accumulated metals

Metals vary in their toxicity and metal ions may be classified into three groups (Fig. 10.2). Class A metal ions are those which bind readily to carboxylic acids or other oxygen-binding sites, but not to nitrogen and sulphur centres on the cell wall or within the cell. These metals are usually non-toxic unless in very high concentrations. Class B metals bind most strongly in living organisms to nitrogen and sulphur centres, which include the amino and sulphydryl groups of enzymes. This results in interference with enzyme function, which is one of the reasons why class B metals are highly toxic. The third group, the borderline metal ions, may bind to either oxygen or nitrogen-sulphur centres depending on the particular metal and its concentration. These metals are toxic when they are present in excessive amounts (Nieboer & Richardson 1980). It is also seen from Fig. 10.2 that the macro-elements

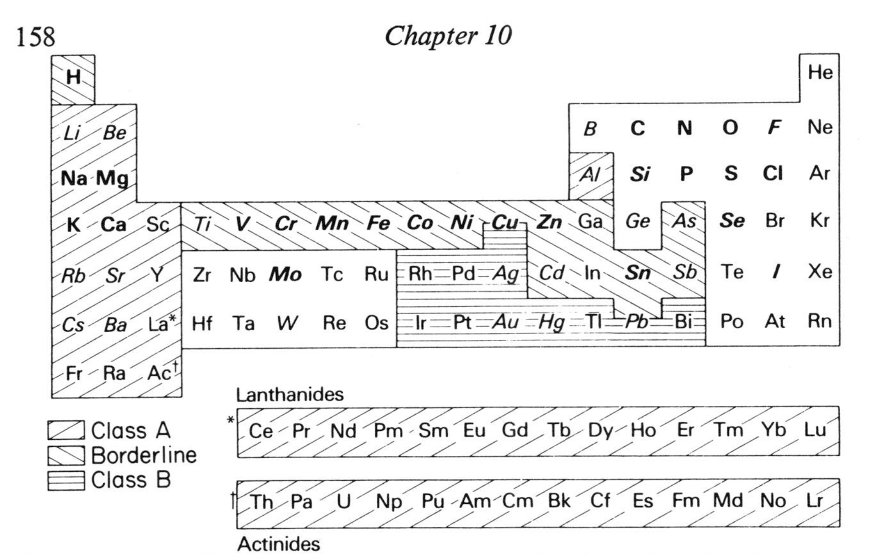

Fig. 10.2 The periodic table of elements showing the disposition of the class A, borderline and class B metal and metalloid ions. Elements designated as essential to mammals are given in bold type (macronutrients) or bold italic (micronutrients) while those that are possibly required are given in light italics (Nieboer & Richardson 1980).

required by living organisms are all class A metals, the essential microelements are borderline metals, while class B metals are strictly avoided as participants in normal metabolism.

Among the borderline ions there is a range of toxicity, but among the most toxic is lead. This metal is a serious environmental pollutant through the use of lead tetraethyl as a petrol additive. If spores of *Funaria hygrometrica* are cultured on media containing lead tetraethyl, germination of the spores is delayed from 2 days by 0·7 ppm and up to 12 days with 66 ppm. Even higher levels (130 ppm) result in 90% failure to germinate and those spores that do, form short protonemata which do not grow. At concentrations of 15 ppm, the protonemata, instead of being elongated, are deformed, branch frequently, exhibit slow growth and contain fewer than normal chloroplasts. These changes are thought to be due to the lead affecting mitotic division of the nuclei within the moss cells (Krupinska 1976). Further evidence of the harmful effects of lead on cell nuclei is derived from electron microscope studies on mosses collected from sites subject to motor vehicle emissions or lead contaminated soil. The nuclei in older leaves are found to contain electron dense precipitates which are considered to be a sign of lead poisoning in both plants and animals (Gullvåg *et al.* 1974). Cultures of *Hylocomium splendens* and *Rhytidiadelphus squarrosus* watered once a

Fig. 10.3 *Grimmia doniana* a tuft forming (acrocarpous) moss of a pollution sensitive genus. This species is found on siliceous rocks at moderate to high altitude (photo: M.C.F. Proctor).

day with a 1000 ppm lead acetate solution were found to discolour within a few days. Treatment with lower dosages (100 ppm or more) in the case of *Hylocomium splendens* showed that lead was bound extracellularly to the cell wall and not intracellularly. In *Rhytidiadelphus squarrosus* which has a thinner wall, some lead was in the cell wall but electron dense vesicles were also found in the cytoplasm. It is thought that the vesicles may be part of a detoxification system and, by emptying into the cell vacuole, maintain low lead levels in the cytoplasm, so perhaps preventing lead from getting into the nucleas (Gullvåg *et al.*

1974). Lead is bound mainly extracellularly in an exchangeable form, but some is in an insoluble form, in cells of *Grimmia doniana*, a moss which grows on lead-rich slag around old mines (Buck & Brown 1978; Fig. 10.3). Presumably mosses from such lead-polluted sites (Crundwell 1976) have evolved detoxification mechanisms comparable to that hypothesized above for *Rhytidiadelphus squarrosus* (Fig 10.4).

Fig. 10.4 *Rhytidiadelphus squarrosus*, a moss which grows in chalk grassland, among lawns and golf courses as well as in woodland. It is one of several commonly used in pollution studies (photo: M.C.F. Proctor).

There are few studies on the effects of other metals on mosses. Experiments on the percentage germination of *Funaria hygrometrica* spores cultured on media enriched with metal ions have shown the toxicity sequence: Cu^{2+}, $Cd^{2+} > Zn^{2+}$ (Coombes & Lepp 1974; Lepp & Roberts 1977). Studies on *Sphagnum* also grown in culture indicated the toxicity sequence: As > Hg > Cu. Arsenic apparently interferes with carbohydrate metabolism and development of vacuoles in cells. The concentration of mercury lethal to *Sphagnum* (20 ppm) is lower than that which kills higher plants in short-term experiments. *Sphagnum* cultured in 2 ppm Hg have chloroplasts with fewer than normal grana, while electron-dense lipid material is found in the cytoplasm. The membranes of the mitochondria and of the whole cells are also abnormal. When the moss is grown on similar concentrations of copper, carbohydrate metabolism is affected and large starch grains develop in the chloroplasts, while cell membranes and organelles are again abnormal (Simola 1977). These toxicity sequences might be expected since As and Hg are class B metals, while Cu and Cd are the borderline metals with most class B character. Zn is a borderline metal, which normally exhibits a significant degree of class A character. The metal toxicity sequence in these moss studies is similar to that for flowering plants i.e. Hg > Pb > Cu > Cd > Cr > Ni > Zn (Nieboer & Richardson 1980).

Both metals and toxic gases, such as SO_2, are often emitted from industrial processes. Experiments on lichens have shown that class A metals, such as magnesium and calcium, tend to protect a plant against the stress imposed by SO_2 exposure (Nieboer *et al.* 1979; Richardson *et al.* 1979a). Calcium also plays an important role, enabling mosses to resist the stress of salinity (Bates 1976).

3 REFERENCES

Barkman J.J. (1969) The influence of air pollution on bryophytes and lichens. In *Air pollution*. Proceedings of 1st European Congress on the influence of air pollution on plants and animals, pp. 197–209. Wageningen.

Bates J.W. (1976) Cell permeability and regulation on intracellular sodium concentration in a halophytic and glycophytic moss. *New Phytol.* **77,** 15–23.

Bell J.N.B. (1973) The effect of a prolonged low concentration of sulphur dioxide on the growth of two moss species. *J. Bryol.* **7,** 444–5.

Brown D.H. & Buck G.W. (1978) Cation contents of acrocarpous and pleurocarpous mosses growing on a strontium-rich substrate. *J. Bryol.* **10,** 199–209.

BUCK G.W. & BROWN D.H. (1978) Cation analysis of bryophytes; the significance of water content and ion location. *Bryophytorum Bibliotheca*, **13,** 735–50.

CLYMO R.S. (1963) Ion exchange in *Sphagnum* and its relation to bog ecology. *Ann. Bot. (Lond.)*, **27,** 309–24.

COMEAU G. & LEBLANC F. (1971) Influence de l'ozone et de l'anhydride sulfureux sur la régénération des feuilles de *Funaria hygrometrica* Hedw. *Nat. Can. (Que.)*, **98,** 347–58.

COOMBES A.J. & LEPP N.W. (1974) The effect of Cu and Zn on the growth of *Marchantia polymorpha* and *Funaria hygrometrica. Bryologist*, **77,** 447–52.

CRUNDWELL A.C. (1976) *Ditrichum plumbicola*, a new species from lead-mine waste. *J. Bryol.* **9,** 167–9.

FERGUSON P., LEE J.A. & BELL J.N.B. (1978) Effects of sulphur pollutants on the growth of *Sphagnum* species. *Environ. Pollut.* **16,** 151–62.

FERGUSON P & LEE J.A. (1979) The effects of bisulphite and sulphate upon photosynthesis in *Sphagnum. New Phytol.* **82,** 703–12.

GILBERT O.L. (1968) Bryophytes as indicators of air pollution in the Tyne valley. *New Phytol.* **67,** 15–30.

GILBERT O.L. (1970) Further studies on the effects of sulphur dioxide on lichens and bryophytes. *New Phytol.* **69,** 605–27.

GILBERT O.L. (1971) The effect of airborne fluorides on lichens. *Lichenologist*, **5,** 26–32.

GULLVÅG B.M., SKAAR H. & OPHUS E.M. (1974) An ultrastructural study of lead accumulation within leaves of *Rhytidiadelphus squarrosus* (Hedw.) Warnst. *J. Bryol.* **8,** 117–22.

HAWKSWORTH D.L. & ROSE F. (1976) *Lichens as Pollution Monitors.* Studies in Biology no. 66. Edward Arnold, London.

INGLIS F. & HILL D.J. (1974) The effect of sulphite and fluoride on carbon dioxide uptake by mosses in the light. *New Phytol.* **73,** 1207–13.

KRUPINSKA I. (1976) Influence of lead tetraethyl on the growth of *Funaria hygrometrica* L. and *Marchantia polymorpha* L. *Acta Soc. Bot. Poloniae*, **45,** 421–8.

LEBLANC F. & RAO D.N. (1974) A review of the literature on bryophytes with respect to air pollution. *Bull. Soc. Bot. Fr. Colloque Bryol.* **121,** 237–55.

LEPP N.W. & ROBERTS M.J. (1977) Some effects of cadmium on growth of bryophytes. *Bryologist*, **80,** 533–6.

NASH T.H. & NASH E.H. (1974) Sensitivity of mosses to sulphur dioxide. *Oecologia (Berl.)*, **17,** 257–63.

NIEBOER E., RICHARDSON D.H.S., PUCKETT K.J. & TOMASSINI F.D. (1976) The phytotoxicity of sulphur dioxide in relation to measurable responses in lichens. In *Effects of Air Pollutants on Plants*, ed. Mansfield T.A. Society for Experimental Biology, seminar series volume 1, pp. 61–85, Cambridge University Press.

NIEBOER E., TOMASSINI F.D., PUCKETT K.J. & RICHARDSON D.H.S. (1977) A model for the relationship between gaseous and aqueous concentrations of sulphur dioxide in lichen exposure studies. *New Phytol.* **79,** 157–62.

NIEBOER E., RICHARDSON D.H.S. & TOMASSINI F.D. (1978) Mineral uptake and release by lichens: an overview. *Bryologist*, **81,** 226–46.

NIEBOER E., RICHARDSON D.H.S., LAVOIE P. & PADOVAN D. (1979) The role of metal-ion binding in modifying the toxic effects of sulphur dioxide on the lichens *Umbilicaria muhlenbergii* I. Potassium efflux studies. *New Phytol.* **82,** 621–32.

NIEBOER E. & RICHARDSON D.H.S. (1980) The replacement of the nondescript term 'heavy metals' by a biologically and chemically significant classification of metal ions. *Environ. pollut. (ser. B)*, **1,** 3–26.

RICHARDSON D.H.S., NIEBOER E., LAVOIE P. & PADOVAN D. (1979a) The role of metal-ion binding in modifying the toxic effects of sulphur dioxide on the lichen *Umbilicaria muhlenbergii* II. ^{14}C-fixation studies. *New Phytol.* **82,** 633–43.

Richardson D.H.S., Nieboer E., Beckett P.J., Boileau L., Lavoie P. & Padovan D. (1979b) *The Levels of Uranium and Other Elements in Lichens and Mosses Growing in the Elliot Lake and Agnew Lake areas, Ontario, Canada.* Report to Ontario Ministry of Environment, Government of Ontario.

Richardson D.H.S., Beckett P.J. & Nieboer E. (1980) Nickel in lichens, bryophytes, fungi and algae. In *Biogeochemistry of Nickel*, ed. Nriagu J.O. John Wiley, New York.

Roberts B.A., Thompson L.K. & Sidhu, S.S. (1979) Terrestrial bryophytes as indicators of fluoride emission from a phosphorous plant, Long Harbour, Newfoundland, Canada. *Cal. J. Bot.* **57,** 1583–90.

Rose F. & Wallace E.C. (1974) Changes in the bryophyte flora of Britain. In *The Changing Flora and Fauna of Britain*, ed. Hawksworth D.L. pp. 27–46. Academic Press, London.

Schiff J.A. & Hodson R.C. (1973) The metabolism of sulfate. *Annu. Rev. Plant Physiol.* **24,** 381–414.

Simola L.K. (1977) Growth and ultrastructure of *Sphagnum fimbriatum* cultured with arsenate, fluoride, mercury and copper ions. *J. Hattori Bot. Lab.* **43,** 365–77.

Tomassini F.D., Lavoie P., Puckett K.J., Nieboer E. & Richardson D.H.S. (1977) The effect of time of exposure to sulphur dioxide on potassium loss from and photosynthesis in the lichen, *Cladina rangiferina* (L.) Harm. *New Phytol.* **79,** 147–55.

Türk R. & Wirth V. (1975) Über die SO_2-Empfindlichkeit einiger Moose. *Bryologist,* **78,** 187–93.

Winner W.E. & Bewley J.D. (1978) Contrasts between bryophyte and vascular plant synecological responses in an SO_2-stressed white spruce association in Central Alberta. *Oecologia* (*Berl.*), **33,** 311–25.

Ziegler I. (1975) The effect of SO_2 pollution on plant metabolism. *Residue Rev.* **56,** 79–105.

Ziegler I. & Hampp R. (1977) Control of $^{35}SO_4^{2-}$ and $^{35}SO_3^{2-}$ incorporation into spinach chloroplasts during photosynthetic CO_2 fixation. *Planta* (*Berl.*), **137,** 303–7.

CHAPTER 11
MONITORING AND GEOBOTANICAL PROSPECTING

1 POLLUTION MONITORING

Mosses have great potential for monitoring the levels of air pollution around urban and industrial centres or even over entire countries. Lichens are of equal value (Hawksworth & Rose 1976; Nieboer & Richardson 1981a). The principal ways in which mosses can be used to assess pollution are: 1 to compare the distribution of mosses in areas around a pollution source with that found in neighbouring unpolluted sites; or 2 to analyse collected samples for the levels of particular elements thought to be emitted.

1.1 Distribution Studies

Such studies are most useful for examining the effects of gaseous pollutants, such as SO_2 or HF, and usually involve investigation of the epiphytic mosses growing on the bark of particular tree species. Based on the presence of particular mosses or lichens, it is possible to devise simple scales which indicate whether a site is polluted. A more sophisticated but time-consuming approach used in Canada, U.S.S.R. and elsewhere, involves calculating an index of atmospheric purity (I.A.P.). The I.A.P. method assesses the degree of air pollution using the formula:

$$\text{I.A.P.} = \sum_{i=1}^{n} (Q \times f)/10$$

where i is a running species number with values 1 to n with n being the total number of epiphytic species present at a particular site; Q is the ecological index of a given species expressed by the average number of other epiphytes occurring with it at all the investigated sites; and f is the frequency-coverage of each species at the site. The frequency is the proportion of trees on which an epiphytic species is physically present at a given site and the coverage is the surface area of the species at the site. Both frequency and cover are assessed as percentages during data collection. From the data frequency and cover data, f is assigned a numerical value on a scale of 1 to 5 for each species. A value of 1 represents a species that is very rare and has a very low coverage and 5 represent a species which is very frequent with a very high coverage.

I.A.P. values are calculated for each site and plotted on a map of the area under study. Zones are delimited by I.A.P. values which fall in the ranges 1–24, 25–49; 50–74 and > 74, and these are drawn by isometric lines. The I.A.P. method enables the determination of areas of high and low atmospheric purity around pollution sources; the results of one such study around a copper mining and smelting area in Quebec are shown in Fig. 11.1 (LeBlanc *et al.* 1974). Sulphur dioxide and metals are emitted by this industrial complex. By relating the I.A.P. values of particular sites for which there is quantitative data on SO_2 levels (from mechanical monitors) with I.A.P. values for other investigated sites, it is possible to build up a picture of the degree of contamination over the whole area under study.

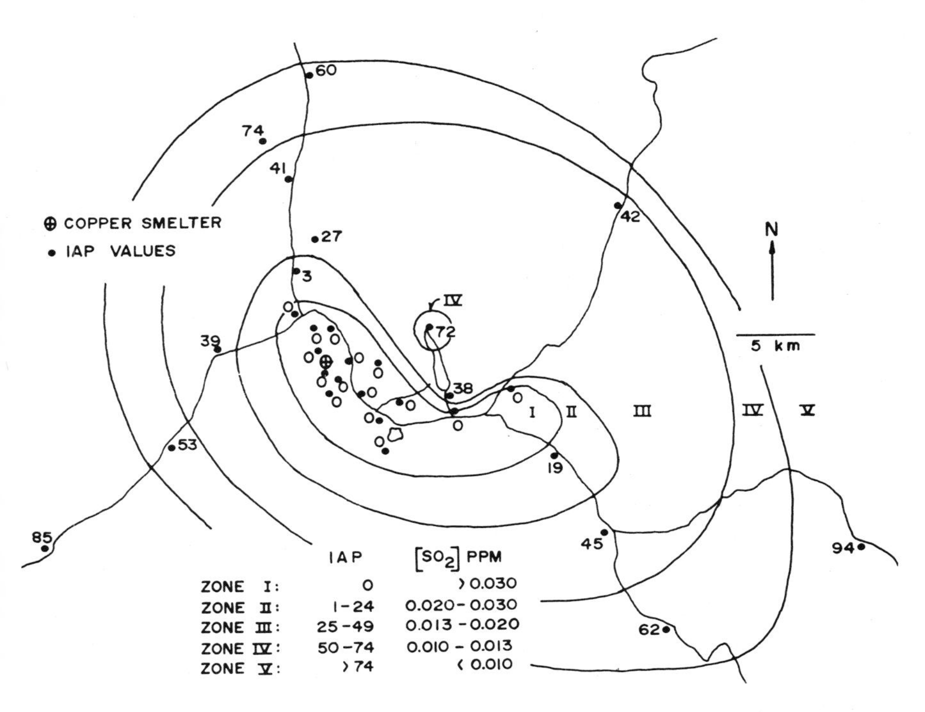

Fig. 11.1 I.A.P. zones (index of atmospheric purity) around the copper mining and smelting centre at Murdochville, Quebec, Canada. These were established by studies on mosses (LeBlanc *et al.* 1974).

1.2 Analytical Studies

By analysing the elements present in mosses it is possible to assess the metal contamination of an area around urban or industrial sites relatively quickly and cheaply. Plants used as monitors for such a purpose should fulfil the following criteria (Laaksovirta & Olkaoknen 1977). 1 The plant should readily accumulate airborne pollutants; the lower the air pollution level which results in elevated pollution contents, the greater is the plant's value as an indicator. 2 The pollutant levels in the plant should vary as a function of distance from the pollution source to permit reliable prediction of the dilution of atmospheric pollutants. 3 The natural content of the pollutant substances in the plant should be fairly constant.

Both peat mosses, true mosses and lichens have been found to meet these criteria rather well (Winner *et al.* 1978) and are of recognized value for determining the zones of influence of industrial pollution, etc. on the surrounding environment.

(a) *Peat Mosses*

Naturally-occurring *Sphagnum*

The ion exchange properties of *Sphagnum* spp. mentioned in the previous chapter, make them most useful in pollution monitoring. In mires, there are frequently hummocks and hollows where particular species of *Sphagnum* grow (*see* also Chapter 12, Section 2.3a). Hummock species such as *Sphagnum fuscum* show metal levels up to five times higher than hollow species such as *Sphagnum balticum* (Pakarinen 1978a). This is due to the higher growth rates and smaller cation exchange capacity of the latter, and it is also possible that metals are more easily leached from the *Sphagnum* in the wet hollows (Pakarinen & Tolonen 1976). Thus, for monitoring purposes samples collected should be of a single species, of comparable age, and from similar habitats (Pakarinen 1978b). *Sphagnum* has been employed, for example, to determine differences in the metal fall-out patterns of Fe, Mn, Zn, Cu, Pb, Cr, Ni, Mo, Cd and Hg over Finland. It was found that samples from the south contained more Pb (17–40 ppm) than those in the less industrial north (3–10 ppm) (Pakarinen & Tolonen 1976).

Studies on peat

It is often useful to know how the fall-out pattern for a particular metal has altered with time, and attempts have been made to do this by analysing peat samples collected at different depths in the deposit. Such investigations are sometimes called palaeoecological studies and they have been used to follow contamination of the environment with lead in a mining district in England (Fig. 11.2). The small peak around A.D. 500 in this figure may represent Roman lead mining. A substantial increase occurs where soot is first obvious, marking the

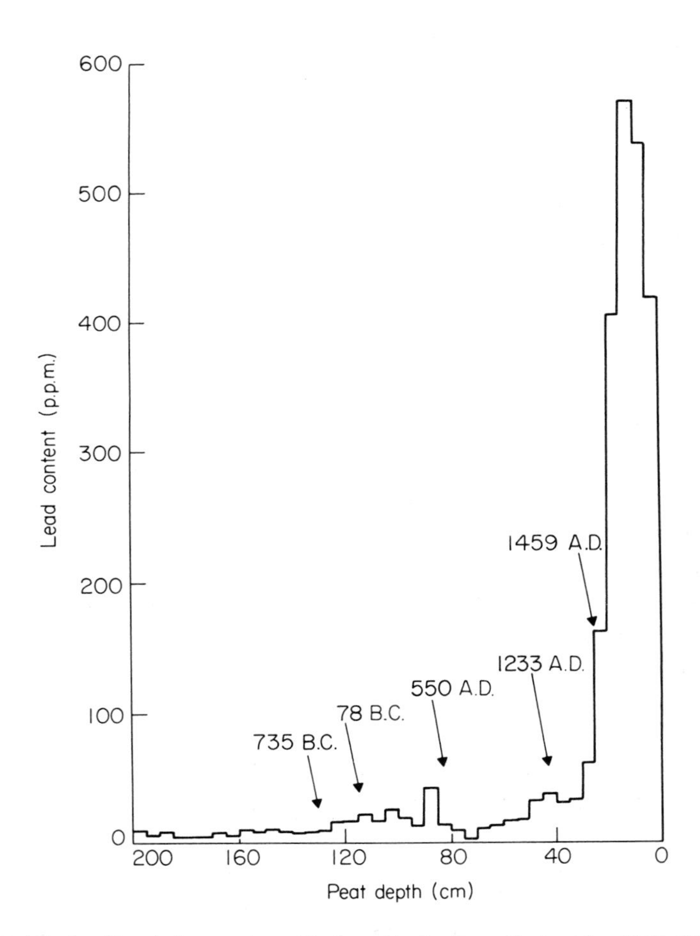

Fig. 11.2 The lead levels in a peat profile from Snake Pass, Derbyshire, U.K. The dates shown are radiocarbon datings (Lee & Tallis 1973).

beginning of industrialization around A.D. 1460. The peak concentrations in the late 19th century are probably related to lead smelting, which was actively occurring in the area at the time (Lee & Tallis 1973).

Recent studies suggest that while man's past activities may be investigated by analysing the humified peat deposits in England (where the ground is not frozen for long periods in winter) elsewhere, e.g. Scandinavia, downward leaching of metals in the less decomposed peat is a problem (Pakarinen & Tolonen 1977). Even in England, some downward movement of metals occurs in the peat profile, as has been shown by investigating concentrations of radionuclides derived from atmospheric nuclear bomb resting. Lead as ^{210}Pb seems to be the least mobile metal followed by ^{239}Pu, with ^{137}Cs leaching fastest. Rapid Cs movement would be expected since monovalent ions are easily displaced from ion exchange sites. However the ^{137}Cs distribution is further complicated by the fact that it is actively taken up by the living *Sphagnum* plants in the upper levels of the bog in mistake for K (Oldfield *et al.* 1979).

Moss bags

There is often a problem of finding mires with healthy *Sphagnum* populations close to the industrial complex or town being studied. This can be solved by the use of moss bags. The technique involves making a flat bag about 10 cm square of nylon net, filled with 2·5 g of clean, acid washed *Sphagnum*. The acid treatment displaces metal ions from the moss so that the binding sites are protonated and in a form best able to exchange with any metal ions in the rainwater. The bags which will also intercept fine dust particles can be suspended from posts of standard height (or tied to trees) at various distances from the pollution source (Roberts 1972; Goodman *et al.* 1974). *Sphagnum acutifolium* has proved to be the most efficient collector of metals. If the uptake of a bag containing this moss is assigned a value of 100%, then bags filled with peat or cotton wool accumulate only 43% and 35% respectively (Mäkinen 1977). Metal retention by moss bags has been found to be directly proportional to exposure time and is linear for Co, Ni, Cu, Zn, Cd and Pb, for up to ten weeks (Goodman *et al.* 1974).

Moss bags have proved of value in determining the fall-out pattern of Hg around chlor-alkali plants (Bull *et al.* 1977), of Ni around smelters in South Wales and northern Ontario, Canada (Goodman & Smith 1975; Richardson *et al.* 1980), and of Zn, Cd and Pb around a smelting complex at Avonmouth in England (Little & Martin 1974; Cameron & Nickless 1977). Metal contents in the moss bags may reach very high levels; for example, in the Avonmouth investigation up to 4,800 ppm Pb, 8,200 ppm Zn and 200 ppm Cd were found. Moss bags have also been used for more fundamental research, as in investigations on nutrient cycling in woodland ecosystems (Hall *et al.* 1977).

(b) *True mosses*

Aquatic forms

The potential of aquatic bryophytes for assessing pollution has been realized only recently. *Fontinalis squamosa* absorbs metal ions efficiently from contaminated water and has been used to study the rivers Ystwyth and Clarach in Wales which are contaminated from old lead mine workings. Samples of *Fontinalis squamosa* may contain up to 1300 ppm Zn, 1200 ppm Pb and 40 ppm Cu (McLean & Jones 1975). When samples of the moss were transplanted to polluted sites, the metal content increased within 6 weeks and the plants started to decay within 18 weeks. *Brachythecium rivulare* and *Eurhynchium riparioides* were less tolerant of polluted water, but readily accumulated metals (MacLean & Jones 1975). *Fontinalis antipyretica* has also been used to assess improvements in water quality on the river Ebbw in Wales following installation of a treatment system to deal with effluent from a steel plant. Living *Fontinalis antipyretica* and *Eurhynchium riparioides* were placed in perspex cylinders, wrapped in nylon netting and secured to stakes in the river. Samples collected up to three weeks later were examined for signs of breakdown and decay. An increase in survival time of moss transplants was observed between 1970 and 1973, which corresponded with improvements in water quality (Benson-Evans & Williams 1976).

Terrestrial forms

Studies using terricolous carpet-forming mosses have enabled assessment of metal contamination of the environment on a regional scale (Richardson *et al.* 1980). For example, many samples of *Hylocomium splendens* collected throughout Scandinavia showed that the heaviest deposition of metals such as Pb and Cd occur in south and southwestern Sweden (Tyler 1972). In north eastern U.S.A., metal concentrations in another moss, *Leucobryum glaucum*, decreased with increasing distance from New York City. Data on moss metal content may be used to estimate average deposition rates, and in the latter instance it is estimated that for Pb, Cu, Cd, Zn, Cr and Ni, they amounted to an average of 33, 2·4, 0·1, 14, 1·3 and 2·3 $mg/m^2/yr$ respectively for this area of the U.S.A. (Groet 1976). On a slightly smaller scale, the contamination of the National Parks of Poland has been investigated by analysing samples of *Pleurozium schreberi* for nine potentially toxic

Fig. 11.3 *Pleurozium schreberi*, a moss much used in pollution studies which is abundant beneath lowland heaths, and in open types of montane and boreal forest (photo: M.C.F. Proctor).

metals, and it was concluded that all parks were seriously contaminated from industrial centres (Fig. 11.3; Grodzinska 1978).

True mosses are also valuable for monitoring where the pollution source is localized. In this case, samples are collected at increasing distances from the source along transects radiating out from it. The most useful mosses are again the ectohydric carpet-forming types such as *Hypnum cupressiforme*, *Pleurozium schreberi* and *Hylocomium splendens* which absorb metals over their entire surface and are little influenced by variations in substrate mineralization. Close to the pollution source such mosses accumulate high levels of any metals being emitted while in control areas, at a distance, very low levels are found (Nieboer *et al.* 1981b). This is illustrated in Table 11.1 by data for

Table 11.1 The average metal content of mosses collected in a polluted zone (see Fig. 11.1) close to the Murdochville Copper mine area, Quebec as compared with the metal content of samples growing at distance from the industrial complex (Zone V, see Fig. 11.1)

Species	Area	Average metal content (ppm dry weight) Pb	Cd	As	Zn	Cu	S	F	Average SO_2 concentration in the air (ppm)
Hylocomium splendens	Polluted: I.A.P. Zone I	17,000	8	4	300	720	1600	7	0.054
	Relatively unpolluted I.A.P. Zone V	200	2	4	130	350	1000	7	0.010
Pleurozium schreberi	Polluted I.A.P. Zone I	700	3	2	230	750	950	5	0.054
	Relatively unpolluted I.A.P. Zone V	60	1	3	90	240	750	7	0.010

Modified from LeBlanc *et al.* (1974)

Hylocomium splendens and *Pleurozium schreberi* that summarizes the extensive study around a copper-mining centre in Quebec (mentioned earlier in Section 1.1; Leblanc *et al.* 1974). *Hypnum cupressiforme* has been used to monitor fall-out of Hg around a chlor-alkali plant in Scandinavia, of Ag in a silver mining area of New Zealand, of Be

Fig. 11.4 *Hypnum cupressiforme*, a species typical of grassy heaths and woodlands as well as tree trunks and calcarious grassland. Its wide occurrence makes it suitable for pollution studies (photo: M.C.F. Proctor).

near electricity power stations in Poland, and of the more common metals around an ironworks in England (Fig. 11.4; Ellison *et al.* 1976; Ward *et al.* 1977; Wallin 1976; Sarosiek *et al.* 1978).

Epiphytic as well as terricolous mosses are sometimes of value in pollution monitoring. Samples of such mosses were collected from marked trees in 1951 and preserved in the Herbarium at Copenhagen, Denmark. The same species, *Homalothecium sericeum*, *Neckera*

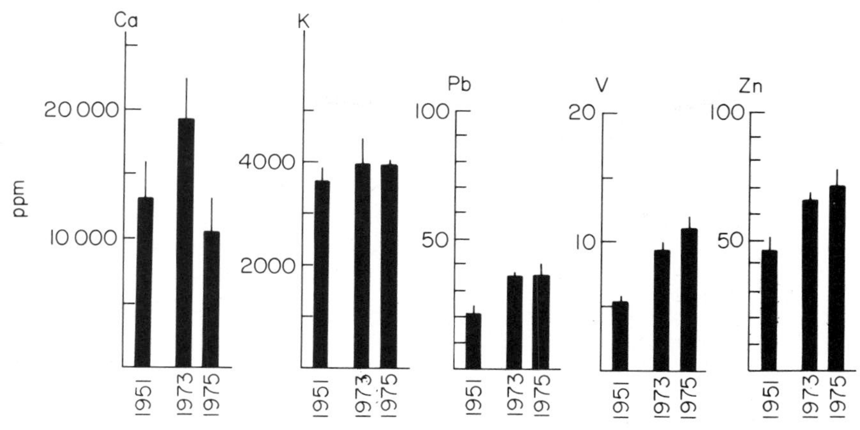

Fig. 11.5 The change in the metal contents of mosses in rural Denmark from 1951–1975. The contents are as ppm dry weight. Values from *Homalothecium sericeum*, *Hypnum cupressiforme*, *Isothecium myosuroides*, *Isothecium myurum* and *Neckera complanata* have been averaged. Vertical lines indicate the upper limits of standard errors (Rasmussen 1977).

complanata, *Isothecium* spp. and *Hypnum cupressiforme* were collected from the same trees again in 1973 and 1975 and analysed for 13 different metals. The content of the macronutrients Ca, K and Mg had not increased over the 14 year period. However, Cr, Cu, Fe, Pb, V and Zn levels had gone up by an average of 60%, with only Cd staying constant (Fig. 11.5; Rasmussen 1977). These epiphytic mosses were collected from rural Denmark so that the rises in mineral content are a reflection of increasing contamination of the whole environment with potentially toxic metals from industrial processes and the burning of fossil fuels.

1.3 Pollution control

Metals and particularly Cd are major factors limiting the use of sewage sludge as a fertilizer on land. The principal use of Cd in the U.S.A. is for electroplating; it is removed from the factory effluent by precipitating the Cd, through adjustment of the pH, followed by sedimentation and perhaps sand filtration. An efficient system of this type results in an effluent with less than 10 μg/l of Cd. Further removal of the Cd may involve adsorption onto starch xanthate, bark waste, activated carbon, or even discarded vehicle tyres.

Recently, metal removal systems have been designed which use unhumified *Sphagnum* peat moss. A system using peat can treat up to

91,000 l (20,000 gallons) of water per day and the waste peat can be buried or burned (Fig. 11.6). *Sphagnum* peat has the advantage that it is very cheap (but see Chapter 12, Section 3.2c for consideration of conservation). It can also accept wide variations in effluent composition and removes a broad spectrum of pollutants (Coupal & Lalancette 1976). A further development for the treatment of Cd suggests passing Cd containing electroplating wastes through columns packed with *Sphagnum* peat moss; emanating water will contain less than 3 μg/l of Cd. In some cases, pretreatment of the *Sphagnum* peat moss with calcium carbonate can be an advantage and it is claimed that *Sphagnum* peat systems have potential for removing contaminants from both industrial and municipal effluents (Poots *et al.* 1976; Chaney & Hundemann 1979).

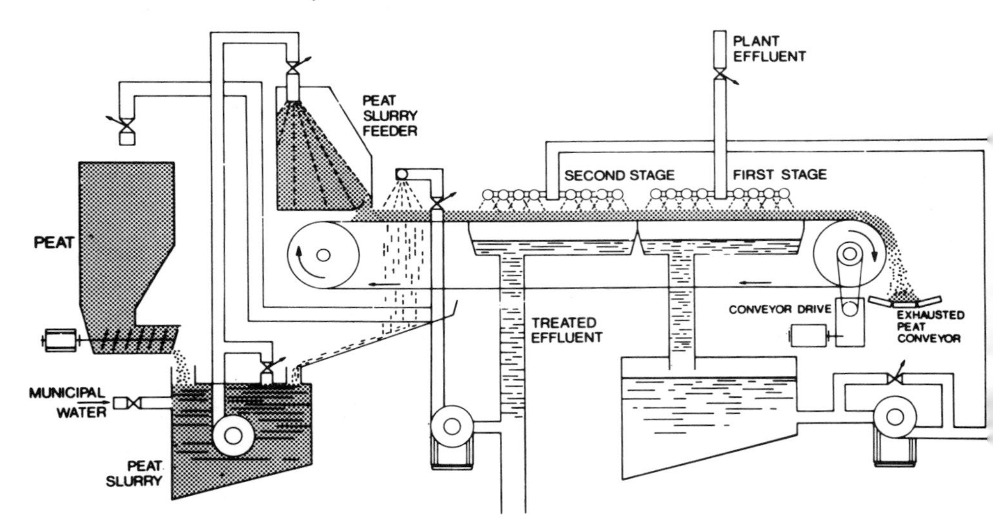

Fig. 11.6 A flow chart for an industrial process employing *Sphagnum* peat for the removal of effluent contaminants (Coupal & Lalancette 1976).

2 GEOBOTANICAL PROSPECTING

A few flowering plants are virtually restricted in their occurrence to soils rich in particular metals and in addition, the metals are often accumulated. By studying the distribution of such plants or analysing them for their metal contents, it is possible to discover new mineral deposits. In northern latitudes where permafrost may hamper the search for underground minerals, investigation of these plants may be particularly useful

(Warren & Delavault 1955). The technique, which may also be applied to mosses, is known as geobotanical prospecting and may even include surveys of herbarium specimens collected by botanists on expeditions in the past (Cannon 1960; Brooks 1979).

2.1 The copper mosses

A number of mosses are so often associated with copper-rich rocks that they have become known as the 'copper mosses'. In one study, a copper moss collected by an amateur from a small mountain in Lapland was presented to an analyst who 'gave a start and rather confused said, "we have just found copper there " ' (Persson 1956). The best known of the copper mosses is *Mielichhoferia elongata* which, in England, occurs as extensive velvety cushions attached to small rock outcrops containing copper sulphides that are kept moist by water seepage on the steep slopes (Fig. 11.7). The moss grows in areas whose aspect ensures that they are seldom exposed to direct sunlight. Physical weathering and microbial activity in these rocks results in the production of sulphuric acid and copper ions; it will be realized, that this is a toxic combination. The copper mosses growing in this habitat accumulate high levels of

Fig. 11.7 *Mielichhoferia elongata* a moss which is often an indicator of copper rich substrates (photo: J.G. Duckett).

copper (Table 11.2), so that studies on both the distribution and the metal content of the mosses may be of value in geobotanical prospecting (Persson 1956; Shacklette 1967).

Table 11.2 The copper content of copper mosses as compared with other mosses

Type	Species	Average vale of: Cu (ppm)	pH of substrate	Locality
Known copper mosses	*Mielichhoferia elongata*	160	3·5	Europe
	Dryptodon atratus	60	3·8	Europe
	Merceya ligulata	95	3·4	Europe, Asia
Other mosses	Other *Mielichhoferia* spp.	30	4·6	Africa, Asia, South America
	Other mosses	35	5·1	Europe

Adapted from Perrson (1956)

The question as to whether copper mosses require large amounts of copper to grow or perhaps only need a sulphur-containing acid and swampy environment has been debated for many years (Schatz 1955; Cockram 1970; Wilkins 1977). It seems likely that copper mosses are simply rather poor at competing with other pioneer mosses but can tolerate both acidity and copper ions. Such a combination is too toxic for other mosses. For example, it has been found that 5×10^{-3} mM solutions of Cu^{2+} are toxic to *Mnium undulatum*, whereas *Mielichhoferia elongata* can withstand 100 mM Cu^{2+} (Url 1956). Thus copper mosses have evolved to occupy a specialized ecological niche but they may also grow on mineral sulphide deposits containing other metals, providing that other mosses are unable to gain a foothold first. Therefore, the presence of a copper moss in an area is by no means an infallible indication of copper deposit and analysis of the moss tissue is needed for confirmation.

A clue as to how the copper mosses survive in their inhospitable environment may be derived from analyses of their elemental content as compared with that of the underlying substrate. The data (Table 11.3)

Table 11.3 The metal content of *Mielichhoferia elongata* as compared with that in the substratum

Sample	Average metal content (ppm dry wt) and standard deviation			
	Ca	Cu	Fe	Pb
Moss	6,600±600	200±80	8,100±1400	590±150
Substrate	2,500±800	370±80	14,400±400	260±80

Adapted from Coker (1970)

suggest that the moss is actually able to exclude much of the copper, and possibly also iron, from its tissues, which exhibit a lower metal content for these elements than the substrate (Coker 1971). Lead seems to be accumulated, but as mentioned earlier, there may be a detoxification mechanism for it. An important observation is that calcium is accumulated in large amounts as compared with the substrate. This element is known to help stabilize cell membranes and provides protection against SO_2 stress (Nieboer *et al.* 1979; as well as playing a role in excluding sodium from seawater-stressed maritime mosses (Bates 1976).

2.2 Other mosses

Mosses indicative of deposits containing metals other than copper have not been studied in detail. *Aerobryopsis longissima* is reported to be an indicator of chromium in New Caledonia (Lee *et al.* 1977) and *Weissia controversa v. densifolia* is associated with lead rich soils in England (Crundwell 1976).

An alternative approach to geobotanical prospecting is to analyse the elemental content of a range of naturally occurring mosses from locations whose overall geology suggests the possibilities of a mineral deposit. High values in the mosses help to define areas that warrant detailed field exploration. For this purpose, the endohydric tuft-forming (acrocarpous mosses) are most suitable, as they tend to absorb metal ions from the base of the moss cushion and hence best reflect substrate enrichment. For example, high Ni levels (*c.* 100 ppm dry wt.) occur in mosses such as *Grimmia apocarpa* from the serpentine rock although the moss is not restricted to this substrate (Shacklette 1965).

Aquatic bryophytes, including *Bryum blandum*, *Dicranella vaginata*, *Fissidens rigidulus* and *Thamnium pandum*, were used in a feasibility study to investigate the value of bryophytes for uranium prospecting in New Zealand. Up to about 3 ppm U, 20 ppm Pb, 2 ppm Cu and 2 ppm Be (dry wt.) were found in mosses where mineralized rocks were present in the watershed. Naturally-occurring living mosses were found to be better indicators than artificial absorbers made of peat sewn into nylon bags and suspended for one week in the rivers. It was concluded that mosses and liverworts afforded a semiquantitative method very suitable for the initial screening of a new area for uranium deposits (Whitehead & Brooks 1969).

3 REFERENCES

BATES J.W. (1976) Cell permeability and regulation of intracellular sodium concentrations in a halophytic and glycophytic moss. *New Phytol.* **77,** 15–23.

BENSON-EVANS K. & WILLIAMS P.F. (1976) Transplanting aquatic bryophytes to assess river pollution. *J. Bryol.* **9,** 81–91.

BROOKS R.R. (1979) Advances in botanical methods of prospecting for minerals. In *Geophysics and Geochemistry in the Search for Metallic Ores*, ed Hood P.J. Geological Survey of Canada, Economic Geology Report bo. 31 pp. 397–410.

BULL K.R., ROBERTS R.D., INSKIP M.J. & GOODMAN G.T. (1977) Mercury concentrations in soil, grass, earthworms and small mammals near an industrial emission source. *Environ. Pollut.* **12,** 135–40.

CAMERON A.J. & NICKLESS G. (1977) Use of mosses as collectors of airborne heavy metals near a smelting complex. *Environ. Pollut.* **7,** 117–25.

CANNON H.L. (1960) Botanical prospecting for ore deposits. *Science* **132,** 591–8.

CHANEY R.L. & HUNDEMANN P.T. (1979) Use of peat moss columns to remove cadmium from wastewaters. *J. Water Pollut. Control. Fed.* **51,** 17–21.

COCKRAM C. (1970) *Mielichhoferia* and *Coscinodon* in Cleveland: a challenge. *Naturalist (Hull)*, 59–62.

COKER P.D. (1971) *Mielichhoferia elongata* (Hornsch.) Hornsch. and *Saelania glaucescens* (Hedw.) Broth, in Scotland. *Trans. Br. Bryol. Soc.* **6,** 317–22.

COUPAL B. & LALANCETTE J. (1976) The treatment of waste water with peat moss. *Water Res.* **10,** 1071–6.

CRUNDWELL A.C. (1976) *Ditrichum plumbicola*, a new species from lead-mine waste. *J. Bryol.* **9,** 167–9.

ELLISON G., NEWHAM J., PINCHIN M.I. & THOMPSON I. (1976) Heavy metal content of moss in the region of Consett (North East England). *Environ. Pollut.* **11,** 167–74.

GOODMAN G.T., SMITH S., PARRY G.D.R. & INSKIP M.J. (1974) The use of moss-bags as deposition gauges for airborne metals, pp. 1–16. *Proc. 41st Conf. Nat. Soc. Clean Air*, Brighton, U.K.

GOODMAN G.T. & SMITH S. (1975) Relative burdens of airborne metals in South Wales. In *Report of a Collaborative Study on Certain Elements in Air, Soil, Plants, Animals and Humans in the Swansea–Neath–Port Talbot area together with a report on a moss-bag study of atmospheric pollution across South Wales*, pp. 337–61. Y Swyddfa Gymreig (Welsh Office), Cardiff.

GRODZINSKA K. (1978) Mosses as bioindicators of heavy metal pollution in Polish national parks. *Water Air Soil Pollut.* **9,** 83–97.

GROET S.S. (1976) Regional and local variations in heavy metal concentrations of bryophytes in the northeastern United States. *Oikos*, **27**, 445–56.

HALL C., HUGHES M.K., LEPP N.W. & DOLLARD G.J. (1977) Cycling of heavy metals in woodland ecosystems. In *International Conference on Heavy Metals in the Environment*, ed. Hutchinson T.C. Toronto, Oct. 27–31, 1975. Symp. Proc. vol. 2 (1) p. 230.

HAWKSWORTH D.L. & ROSE F. (1976) *Lichens as Pollution Monitors*. Studies in biology, no. 66. Edward Arnold, London.

LAAKSOVIRTA K. & OLKKONEN H. (1977) Epiphytic lichen vegetation and element contents of *Hypogymnia physodes* and pine needles examined as indicators of air pollution at Kokkola, W. Finland. *Ann. Bot. Fenn.* **14**, 112–30.

LEBLANC F., ROBITAILLE G. & RAO D.N. (1974) Biological response of lichens and bryophytes to environmental pollution in the Murdochville copper mine area, Quebec. *J. Hattori Bot. Lab.* **38**, 405–33.

LEE J.A. & TALLIS J.H. (1973) Regional and historical aspects of lead pollution in Britain. *Nature* (*Lond.*), **245**, 216–18.

LEE J., BROOKS R.R. & REEVES R.D. (1977) Chromium-accumulating bryophyte from New Caledonia. *Bryologist*, **80**, 203–5.

LITTLE P. & MARTIN M.H. (1974) Biological monitoring of heavy metal pollution. *Environ. Pollut.* **6**, 1–19.

MAKINEN A. (1977) Moss- and peat-bags in air pollution monitoring. *Suo*, **28**, 79–88.

MCLEAN R.O. & JONES A.K. (1975) Studies of tolerance to heavy metals in the flora of the rivers Ystwyth and Clarach, Wales. *Freshwater Biol.* **5**, 431–44.

NIEBOER E., RICHARDSON D.H.S., LAVOIE P. & PADOVAN D. (1979) The role of metal-ion binding in modifying the toxic effects of sulphur dioxide on the lichens *Umbilicaria muhlenbergii*. I. Potassium efflux studies, *New Phytol.* **82**, 621–32.

NIEBOER E. & RICHARDSON D.H.S. (1981a) Lichens as monitors of atmospheric deposition. In *Atmospheric Inputs of Pollutants to Natural Waters*. ed Eisenrich S.J. Ann Arbor Science Publishing, Michigan.

NIEBOER E., RICHARDSON D.H.S., BOILEAU L., BECKETT P.J.B. & HALLMAN E.D. (1981b) Definition of the sphere of influence of mining activities at Elliot Lake, Ontario by assessment of the levels of uranium and other elements in Lichens and mosses. proc. 1st Technical Transfer seminar, Ontario Ministry of the Environment, Toronto, Ontario (November 25th, 1980).

OLDFIELD F., APPLEBY P.G., CAMBRAY R.S., BARBER K.E., BATTARBEE R.W., PEARSON G.R. & WILLIAMS J.M. (1979) ^{210}Pb, ^{137}Cs and ^{239}Pu profiles in ombrotrophic peat. *Oikos*, **33**, 40–5.

PAKARINEN P. & TOLONEN K. (1976) Regional survey of heavy metals in peat mosses (*Sphagnum*). *Ambio*, **5**, 38–40.

PAKARINEN P. & TOLONEN K. (1977) Distribution of lead in *Sphagnum fuscum* profiles in Finland. *Oikos*, **28**, 69–73.

PAKARINEN P. (1978a) Distribution of heavy metals in *Sphagnum* layer of bog hummocks and hollows. *Ann. Bot. Fenn.* **15**, 287–92.

PAKARINEN P. (1978b) Element contents of *Sphagna*: variation and its sources. *Bryophytorum Bibliotheca*, **13**, 751–62.

PERSSON H. (1956) Studies in 'copper mosses' *J. Hattori Bot. Lab.* **17**, 1–18.

POOTS V.J.P., MCKAY G. & HEALY J.J. (1976) The removal of acid dye from effluent using natural absorbents–I. Peat. *Water Res.* **10**, 1061–6.

RASMUSSEN L. (1977) Epiphytes as indicators of the changes in the background levels of airborne metals from 1951–75. *Environ. Pollut.* **14**, 37–45.

RICHARDSON D.H.S., BECKETT P.J. & NIEBOER E. (1980) Nickel in lichens, bryophytes, fungi and algae. In *Biogeochemistry of Nickel*, ed. Nriagu J. John Wiley, New York.

ROBERTS T.M. (1972) Plants as monitors of airborne metal pollution. *J. Environ. Plann. Pollut. Control*, **1,** 43–54.

SAROSIEK J., KWAPULINSKY J. & BUSZMAN A. (1978) Bryophytes as biological indicators of beryllium. *Bryophytorum Bibliotheca*, **13,** 763–75.

SCHATZ A. (1955) Speculations on the ecology and photosynthesis of the 'copper mosses'. *Bryologist*, **58,** 113–20.

SHACKLETTE H.T. (1965) Bryophytes associated with mineral deposits in Alaska. *U.S. Geol. Surv. Bull.* **1198–C,** 1–18.

SHACKLETTE H.T. (1967) Copper mosses as indicators of metal concentrations. *U.S. Geol. Surv. Bull.* **1198–G,** 1–18.

TYLER G. (1972) Heavy metals pollute nature, may reduce productivity. *Ambio*, **1,** 52–9.

URL W. (1956) Uber Schwermetall-, zumal Kupferresistenz einiger Moose. *Protoplasma*, **46,** 768–93.

WALLIN T. (1976) Deposition of airborne mercury from six Swedish chlor-alkali plants surveyed by moss analysis. *Environ. Pollut.* **10,** 101–14.

WARD N.I., BROOKS R.R. & ROBERTS E. (1977) Heavy metals in some New Zealand bryophytes. *Bryologist*, **80,** 306–12.

WARREN H.V. & DELAVAULT R.E. (1955) Biogeochemical prospecting in northern latitudes. *Trans. R. Soc. Can. Ser.* III, **49,** 111–15.

WHITEHEAD N.E. & BROOKS R.R. (1969) Aquatic bryophytes as indicators of uranium mineralization. *Bryologist*, **72,** 501–7.

WILKINS P. (1977) Observations on ecology of *Mielichhoferia elongata* and other 'copper mosses' in the British Isles. *Bryologist*, **80,** 175–81.

WINNER W.E., BEWLEY J.D., KROUSE H.R. & BROWN M.H. (1978) Stable isotope analysis of SO_2 impact on vegetation. *Oecologia* (*Berl.*), **36,** 351–61.

CHAPTER 12
MOSSES AND MAN

1 TRUE MOSSES

1.1 Archaeological records

(a) *Bedding, stuffing and caulking*

Little use is made of true mosses by modern man and there is no record of granite mosses being of importance at any time although scientifically they are a very interesting group. In the past it is evident that man looked upon true mosses with more favour. Excavations at Vindolanda, the site of a Roman fort in northern England, just south of Hadrians Wall, revealed both artifacts (cloth, leather, writing tablets, etc.) and plant remains in a remarkable state of preservation (Seaward 1976). Apparently the people who dwelt on the site from about A.D. 90 to A.D. 120 covered the floors of some of the buildings with thick layers of dried bracken fern, straw and moss. Eighty-five percent of the moss belonged to two species *Hylocomium splendens* and *Rhytidiadelphus squarrosus*, with the former being the more abundant (Seaward & Williams 1976). Incidentally puffballs were also found in the resulting organic deposit and are thought to have been used to staunch the bleeding of cuts (in soldiers or workers) or as tinder (Watling & Seaward 1976).

At another site in York, England, which could be as late as A.D. 1200, pits were found which had been carefully lined with moss (mainly *Pleurozium schreberi*, *Hypnum cupressiforme* and *Hylocomium splendens*) and were thought to be used for the storage of vegetables and fruit (Seaward & Williams 1976). Evidence suggests that ancient man also packed mosses between the mainly unfashioned wall timbers in their houses both in England and the lake dwellings in Europe (Dickson 1973; Seaward & Williams 1976). A related and very important use of moss was as caulking material between the planks of boats built by both Bronze Age (*c.* 800 B.C.) and Iron Age (*c.* 200 B.C.) man (Dickson

1973). Wads of various mosses, especially *Neckera complanata*, were used though the reason for selection of this species is obscure (Fig. 12.1).

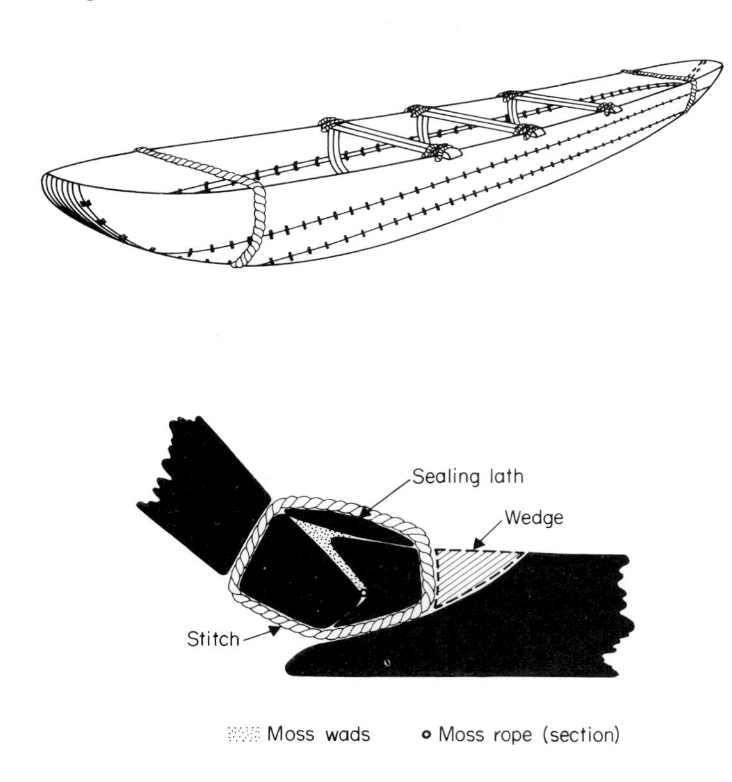

Fig. 12.1 The construction of iron age boats and method of using moss for caulking. The moss wads are *Neckera complanata* and the moss rope is *Polytrichum commune* (Dickson 1973).

(b) *Fibre*

Polytrichum is the first moss to have been illustrated in early herbals and to have an English name (the golden maidenhair moss and later the hairy cap moss). This may have been due to its utility, since the shoots of this moss which grow up to 30 cm long, when once stripped of leaves, form tough pliable strands which can be plaited (Dimbleby 1967). A cap or partly finished basket made of this material was found preserved at a site in Newstead Abbey, Scotland and has been dated around A.D. 90 (Fig. 12.2; Thieret 1956). In addition, this moss and others were

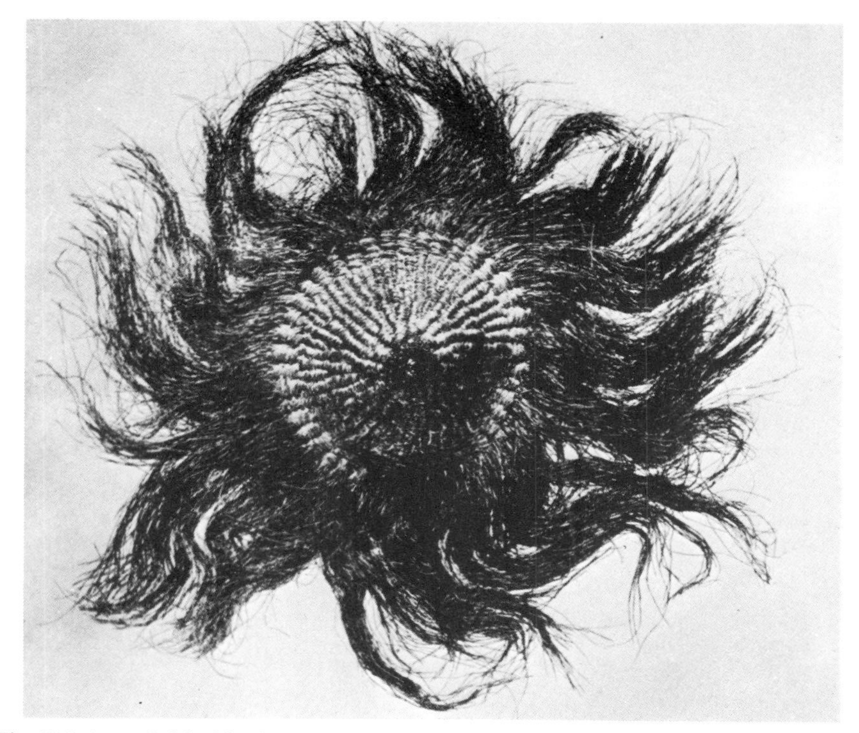

Fig. 12.2 An unfinished basket plaited from the stems of *Polytrichum commune* which was found at Newstead Abbey and dates from the Roman period (Thieret 1956).

central to the techniques of Bronze Age boat builders. The remains of their craft, up to 15 m long and formed of oak planks stitched together, have been excavated in northern England. Ropes made of *Polytrichum commune* were sometimes employed in the construction (Fig. 12.1; Dickson 1973).

1.2 Traditional records

It is not surprising that a number of uses to which ancient man put moss were continued until quite recent times. Thus moss was imported into Holland from Belgium, steeped in tar, and employed for caulking boats at least until after the sixteenth century. In northern Scotland moss was used for this purpose as late as the early nineteenth century (Dickson 1973). *Pleurozium schreberi* was reported by Linnaeus to be the moss collected to block chinks in walls in Scandinavia and elsewhere (Fig.

Fig. 12.3 The frontispiece to the *Moss House* (1822) illustrating the traditional use of mosses for filling chinks in house walls. 'In a few days nothing remained but to procure the coloured glass, to thatch the roof, and to fill the open spaces between the boughs with moss.'

12.3; Anon. 1822). This is still practiced during the construction of log houses in the U.S.S.R. (Seaward & Williams 1976). In Europe quantities of the aquatic moss *Fontinalis antipyretica* were employed for filling spaces between chimneys and walls in order to exclude air and thus prevent the walls from catching fire. Hence the moss derived the specific name *antipyretica*, which literally means 'against fire'. This name has led to the erroneous idea that the moss itself will not catch fire (Bagnall 1889).

In northern England, *Polytrichum* was reported as superior to straw for stuffing mattresses used by less well-off citizens while *Hypnum* (a name derived from the Greek word for sleep) was best for pillows. *Polytrichum* was also used for both bed and bedding by Laplanders. Indeed Linnaeus during his travels in Scandinavia reported that 'I have often made use of it with admiration . . . I could almost imagine that our quilts were but an imitation of it. They fold this bed together tying it with a cord . . . and carry it to the place where they mean to sleep the night following. If it becomes too dry and compressed, its former elasticity can be restored by moisture' (Bagnall 1889). The same moss was sometimes used to weave doormats in northern England, while in the south the foresters of some areas used it to make small brushes. The famous naturalist Gilbert White said that 'when this moss is well combed, dressed and divested of its outer skin, it becomes a beautiful bright chestnut brown and being soft and pliant is very proper for the dusting of beds, curtains, carpets, hangings, etc.' (White 1832).

The Indians of North America also valued mosses. The Chippewa used *Dicranum bonjeanii* as an absorbant and the Cree found a use for another member of this genus, *Dicranum elongatum* for lamp wicks (Thieret 1956). The Gosuite made a kind of paste from *Philonotis*, *Bryum*, *Mnium* and various hypnaceous mosses, and applied it as a poultice to alleviate the pain of burns. Mosses were also used for covering bruises and as padding under splints in the setting of bones (Flowers 1957). Finally the Inuit (Eskimos) of North America made lamp wicks out of *Rhacomitrium lanuginosum*, and were also reported to make a moss bed on which the bodies of the dead were laid for burial (Thieret 1956).

In Victorian times, mosses were much collected in England, cleaned and sold in markets to put around the base of potted plants and for flower arrangements; indeed an instructive and moral tale for children entitled *The Moss Gatherers* was built around this occupation (Anon. 1867). In America around 1900, the ladies of Boston liked to trim their hats and bonnets with imported braids and cords made of moss. The former were constructed of *Pseudoscleropodium purum* while the latter were made of *Neckera crispa* (Clarke 1902). In New York, at the same period, bunches of *Climacium dendroides* and moss roses of *Hylocomium splendens* could be purchased (Britton 1902). The decorative value of mosses is also appreciated by the natives of New Guinea who incorporate considerable quantities of moss in headresses used for ceremonial occasions (Fig. 12.4). In England bryologists were not impressed by the Victorian fashion for using mosses and one wrote 'nor is a bunch of moss

Fig. 12.4 A ceremonial dress from New Guinea, Australasia. The long pendant masses of vegetation on either side of the mask are mosses (photo: G. Argent).

which has died of thirst suitable for trimming a ladies bonnet . . . and it is not improved by being dyed . . . a leather colour or violent blue-green' (Tripp 1874).

The psychological value of mosses should not be overlooked. One has only to consult a dictionary of literary quotations to realize how much these plants have inspired poets in the past. More remarkable perhaps was the response of Mungo Park, the African explorer. He was captured by the Arabs but managed to escape after four months. He faced a journey of 3,000 km through unknown country alone. In addition to suffering hunger and fever, he was robbed of his compass and all but shirt, trousers and hat. He wrote 'I considered my fate as certain and that I had no alternative but to lie down and perish. . . . At this moment, painful as my reflections were, the extraordinary beauty of a small moss in fructification irresistably caught my eye. I mention this to shew from what trifling circumstances the mind will sometimes derive consolation. . . . Can that being (thought I), who planted, watered and brought to perfection, in this obscure part of the world, a thing which appears of so small importance, look with unconcern upon the situation and sufferings of creatures formed after his own image? Surely not! Reflections like these would not allow me to dispair. I started up and disregarding both hunger and fatigue travelled forwards assured that relief was at hand.' He arrived in London safely after a journey of 19 months. Unfortunately when Mungo Park returned to Africa in 1805 he perished, during an attack by natives, in rapids on the river Niger (Park 1799; Crum 1973).

1.3 Contemporary importance

In the last two decades, the value of mosses in pollution monitoring and pollution control has been realized and this forms the subject of a separate chapter.

(a) *Moss cultivation*

Modern society still appreciates mosses for their decorative effect. Two tropical aquatic mosses, *Vesicularia dubyana* and *Glossadelphus zollingeri* are cultured and sold for transplanting into fish tanks to provide oxygen and a more natural appearance (Cook *et al.* 1974).

The cultivation of mosses reaches its greatest refinement in Japan (Ishikawa 1974). There, landscape gardeners carefully select rocks and trees planting mosses in between because they help to induce the desired feeling of tranquillity of mind in a visitor. Although small moss gardens are popular among amateurs, the most famous moss garden is in the

Fig. 12.5 Moss gardens of Japan. (a) A view inside the gates of Saihoji Temple, Kyoto. *Brachythecium buchananii*, the whitish green belt near the path, withstands trampling. (b) Wavy carpets of *Leucobryum neilgherrense* and *Leucobryum bowringii* in the moss gardens. On the stream bank *Polytrichum commune* is dominant (photo: H. Ando).

temple grounds at Saihoji at the foot of Mount Koinzan in Western Kyoto province in central Japan. The temple is known as the Moss Temple and is surrounded by beautiful gardens with thick growths of mosses comprised of some fifty different species (Fig. 12.5). A relatively high water table, high humidity and moderate light intensity ensure that the mosses planted on compacted earth grow well. Undesirable species of moss are weeded out and the cushions are kept free of leaves and debris (Iwatsuki & Kodama 1961). In open sites *Polytrichum* is planted, whilst in moist shaded places *Rhizogonium* and *Mnium* form beautiful soft carpets. Underneath trees, *Leucobryum* and *Bazzania* form undulating hummocky waves of whitish green. *Brachythecium*, which can resist trampling, forms a distinctive belt near paths (Ando 1971). The high regard in which moss gardens are held in Japan may be gauged from the fact that in only three weeks more than 60,000 people visited a special exhibition of mosses at the National Science Museum in Tokyo (Inoue 1972; Ando 1980).

(b) *Orchards and pastures*

The modern practice, for the cultivation of fruit trees such as apples and pears on a commercial scale, is that orchards are either mown to produce short grass or sprayed with herbicide. The latter procedure results in bare ground remaining between the trees and this is colonized by mosses. *Polytrichum* sp., *Atrichum undulatum* and *Ceratodon purpureus* commonly develop on paraquat treated plots in the British Isles whilst *Bryum argenteum* becomes dominant if Simazine is used (Bond 1976). The yield of fruit in orchards treated with herbicides is reported to be increased significantly (20%) and the trees are apparently more vigorous under this regime. The moss is valuable in that it inhibits soil erosion and provides good conditions for machinery passage (Robinson & O'Kennedy 1978). The movement of new varieties of fruit trees between countries has resulted in transcontinental travel of mosses, e.g. *Pseudoscleropodium purum* and the pollution-tolerant lichen *Lecanora conizaeoides*. These plants are common in Europe and have now spread to areas such as the west coast of North America where gardening interests are strong (Crum 1972; Richardson 1975). Movement of mosses in the other direction e.g. *Tortula amplexa* (see Chapter 6) is also known. Mobile dispersal of mosses through man's agency on a more local scale is occasionally reported. Thus an aged, but still functional, Morris Traveller car in England (an estate model with a wooden frame at the rear) had tufts of *Ceratodon* sp., *Tortula muralis*, *Bryum argenteum* and *Grimmia pulvinata* growing on it. All but the first were exhibiting a fine crop of capsules and such instances of colonization of old vehicles are quite common (Anonymoss 1979).

The growth of moss in pasture and lawns is a definite disadvantage as it chokes the more useful grasses and herbs. Man spends considerable effort and money to eradicate mosses, e.g. *Brachythecium* spp. from such situations. Calomel (Hg_2Cl_2) kills established moss rather slowly but has a residual effect inhibiting the germination of moss spores and is therefore best used in autumn. Tar acids give a quick kill but have no residual effect and are useful in spring as a clean-up treatment. A less drastic, ecologically less toxic approach is to spread lawn sand which contains iron sulphate and fertilizers: the former being detrimental to mosses while the latter stimulate herbs and grasses (Frye 1920; Stubbs 1973). The removal of mosses from tile roofs, by scraping or use of toxic chemicals, is often recommended as the plants tend to maintain high moisture levels in the outer layers of the tile and this may lead to flaking or cracking under freezing conditions in winter. Similarly, mosses may be a problem on cedar shingles, a popular roofing material in America, by greatly increasing the rate of decomposition due to the moisture held by the moss (Frye 1920).

Modern urban man is subjected to a high degree of architectural uniformity via the utilization of characterless concrete, glass and tar-macadam. The implementation of clean air acts in many countries has provided the opportunity for lower plants to invade these substrates, sometimes producing unsightly effects but often softening the hard outlines of buildings. *Ceratodon purpureus*, *Bryum argenteum* and *Tortula* spp. are mosses which are commonly able to tolerate the drought and pollution of urban environments (Seaward 1979). It has even been suggested that citizens should encourage the growth of mosses on rocks, walls and other structures by coating them with buttermilk, cooked oatmeal, or a slurry of cow manure (Anon. 1975b).

2 PEAT MOSSES

2.1 FRESH SPHAGNUM

In Britain, the Forestry Commission invites tenders for the privilege of collecting moss from particular areas under its control. Successful applicants have the exclusive right to collect moss, and it is said that the period before Christmas and in the spring is most busy (Anon. 1968). The moss is picked, cleaned and squeezed to remove the bulk of the contained water. It is packed and sent to customers to line hanging baskets, for floral displays, especially in horticultural shows (including the famous Chelsea Flower show), for wreaths, and for protecting the roots of plants being dispatched from nurseries to customers. It is also employed in nurseries to assist aerial rooting (by being wrapped tightly around stems) and in the propagation of rare orchids. Finally, moist *Sphagnum* is used to pack frogs (Bland 1971), worms and some insects for shipment from laboratory supply houses to schools and colleges.

2.2 DRIED SPHAGNUM

(a) *Sphagnum dressings*

The virtue of *Sphagnum* as an absorbent dressing has been known for centuries, being used by country people in the British Isles for treating boils and discharging wounds. *Sphagnum* dressings were first used on a large scale during the Russo–Japanese war, 1904–1905 (Thieret 1956). However, it was in the First World War that their value was widely recognized by both sides. The British used a million *Sphagnum* dressings a month and the Germans probably more. However, American surgeons disliked this type of dressing unless disguised with a thin layer of cotton (Hotson 1921a, b). The advantages of *Sphagnum* dressings included: 1 they absorbed 16–20 times their own dry weight of liquid whereas cotton dressings would only absorb 4–6 times; 2 they were cool and soothing because the moss was porous; a dressing was often found to be comparatively dry 24 hours after an operation in which the wound had bled quite freely; 3 the dressings could be left on for up to 2–3 days which is much longer than cotton dressings so that soldiers could be transported from the field to hospitals without disturbing the wound or exposing it to further infection during dressing changes under primitive conditions; 4 *Sphagnum* itself has mild antiseptic properties not possessed by cotton.

The considerable absorptive powers of *Sphagnum* are related to the anatomy of the plant. In addition to the small green photosynthetic cells, the leaves have large colourless dead cells with many pores. The capacity to hold liquid is also aided by the closely overlapping leaves and pendent branches which clasp the stem so providing many spaces for capillary uptake.

By 1918 when the World War ended, factories had been established for the production of *Sphagnum* dressings in Germany, U.K., U.S.A. and Canada. In North America, the large, robust, yellow-brown species, *Sphagnum papillosum* was considered of most commercial value due to the numerous large pores in the dead cells and greatly differentiated outer layers of the stem which combined to give the greatest absorption. *Sphagnum palustre*, *Sphagnum magellanicum* and *Sphagnum imbricatum* were also used but other species were considered of little value (Nichols 1918 a, b; Hotson 1921 a, b). Clean dry *Sphagnum* was enclosed in a layer of fine muslin, the edges sewn up and the whole then sterilized. Machinery was invented to clean and make the dressings. After hostilities ceased, the use of *Sphagnum* dressings declined and their virtues became forgotten. While the cost of collection of the moss was no doubt a factor, the greater aesthetic appeal of a white cotton dressing to both patients and the medical profession was probably the main reason for the demise of the *Sphagnum* counterpart. During the Second World War, cotton was never sufficiently scarce to warrant the resumption of manufacture of moss dressings (Thieret 1956; Crum 1973).

(b) *Miscellaneous uses*

Dried *Sphagnum* is apparently sold in herb shops in various parts of China where a decoction is said to be effective for the treatment of haemorrhages and eye diseases (Shiu-ying 1945). The Inuit (Eskimos) traditionally used this moss to pad their sealskin boots in lieu of socks, while Linnaeus reported that 'Lapland matrons are well acquainted with this moss. They dry and lay it in their childrens' cradles to supply the place of bed, bolster and every covering, and being changed night and morning it keeps the infant remarkably clean, dry and warm. It is sufficiently soft of itself but the tender mother, not satisfied with this frequently covers the moss with the downy hairs of reindeer, and by that means makes the most delicate nest for the new-born babe' (Crum 1973). It was used for the same purpose by the Indians of Alaska and those around Hudson's Bay, Canada. Finally, during the First World War, large couch pillows stuffed with *Sphagnum* were produced and used to support the limbs of wounded soldiers or make their travel in troop trains more bearable (Hotson 1921a, b).

2.3 PEAT

(a) *Deposition of peat*

The formation of peat results from the accumulation and compression of the remains of plants including mosses, sedges, grasses, reeds and shrubs. Mosses, particularly *Sphagnum*, are the key plants responsible for deep, rapidly accumulated deposits. For example, in the bogs in central Ireland, reed swamp peat deposits 1·5 m in thickness took 6,000 years to accumulate. *Sphagnum* then became dominant on the bogs (about 500 B.C.) and true acid peat began to form over the older deposits and rapidly built up, so that in the period from 500 B.C. until the present, a thickness of some 5 m accumulated which consists largely of *Sphagnum* remains (Anon. 1979b; Walker & West, 1970). The enhanced rate of peat formation results from the capacity of *Sphagnum* to flourish in habitats which are uncommonly low in inorganic ions especially N and P. The concentration in the plant mirrors that in the habitat being only about 0·6 % N and 0·03 % P. In addition the pH which is maintained by the plant through the release of H^+ ions (see Chapter 10) is often below 4. This added to the ability of *Sphagnum* to hold water ensures that acid, waterlogged anaerobic conditions develop beneath the *Sphagnum* plant so inhibiting competition from other plants and microbial activity. Thus the rate of breakdown of both moss and plant litter is reduced and peat accumulates (Clymo 1970, 1978).

The greatest growth of the *Sphagnum* mosses in mires occurs during the summer though a resting phase may be observed if the surface of the bog dries to any extent. In one study in Northern England the estimated net production of dry matter by *Sphagnum papillosum* was estimated as 3 t/ha/yr as compared with 2 t/ha/yr in nearby grassland and 13 t/ha/yr for a pine forest in Southern England (Clymo 1970). Thus growth and production by *Sphagnum* is at least comparable with other plant communities from the same area. The surface of mires is often a complex of hummocks and hollows, each colonized by different species (see Chapter 11, section 1.1a). In some instances hummocks replace hollows in the peat profile, as determined by examining the remains of the mosses, in others they do not. In any case the rate of peat accumulation is in the order of between 20 and 80 cm per 1000 years (Moore & Bellamy 1974).

The area covered by peat bogs, more properly called mires, is vast, especially in the temperate and subarctic regions and is, at a conservative estimate, 1% of the whole world's surface (Clymo 1970). Even in the British Isles, there are some 2 million hectares (6 million acres) of peat having an average depth of about 4 m; of this about half is in Ireland.

(b) *Archaeological and palaeobotanical aspects*

The acid, anaerobic character of most peat deposits ensures that microbial degradation of buried organic materials is extremely slow. Thus organisms or their parts which are trapped in the peat do not decompose and may be excavated at a later date, beautifully preserved. This is dramatically emphasized by the seven hundred or so Iron and Bronze age human corpses which have been recovered from Scandinavian peat bogs (Fig. 12.6). In the best of these, the skin and clothes

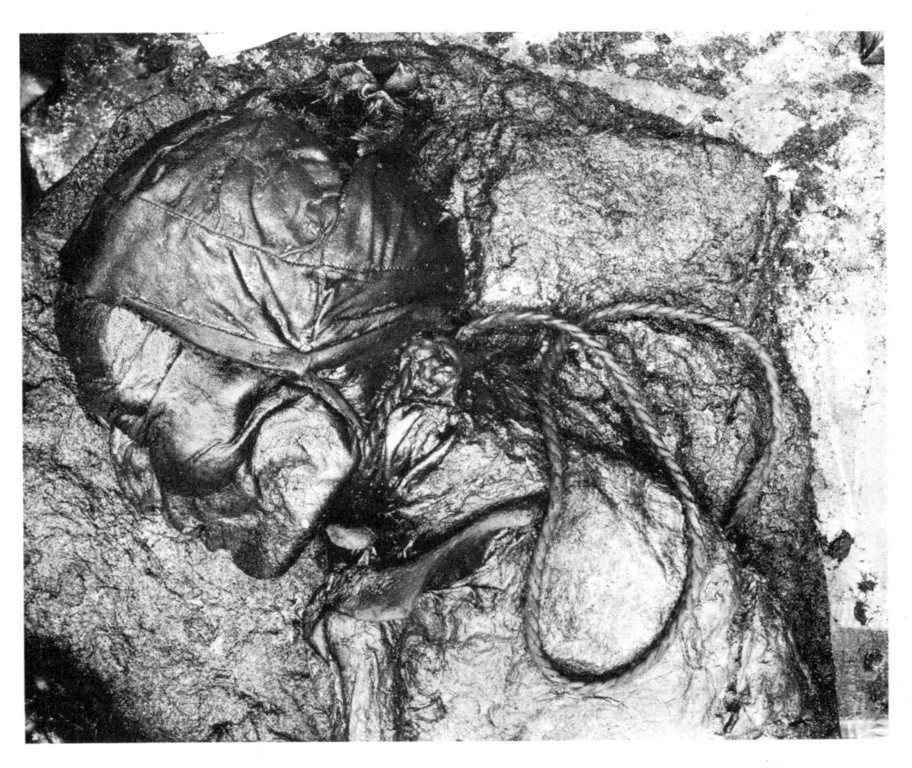

Fig. 12.6 The Tolund man who owes his remarkable state of preservation to burial in a *Sphagnum* mire. Note he still has his cap and the noose with which he was strangled (Glob 1971).

are preserved and even the stomach contents may be analysed to determine the nature of their last meal. Most seem to have been strangled with cords of braided leather as part of a sacrificial ceremony, possibly to a fertility goddess (Glob 1971). Interring people in *Sphagnum* bogs was apparently quite popular in ancient times as Tacitus, the Roman historian writing shortly before A.D. 100, records that in the German tribes 'Traitors and deserters are hanged on trees; cowards, shirkers and the unnaturally vicious are drowned in miry swamps under a cover of wattle hurdles'. There is also a tradition that the body of King Abel, which was buried in Schleswig Cathedral in 1252, would not lie quiet because he murdered his brother. His body was dug up and then reburied in a peat bog near Gottorp with a stake run through it. This prevented his uneasy ghost from causing further disturbance (Bland 1971).

The remains of plants as well as animals may be preserved in *Sphagnum* bogs. Because the different genera (and often species) of flowering plants usually have characteristic pollen grains, it is possible by analysing the pollen from a peat sample to establish what plants were growing and shedding pollen grains around the bog being studied at any given time. The technique is to collect peat samples at varying known depths from the surface of the peat deposit and then digest the peat and examine the pollen grains in each sample with a microscope, a discipline called palynology (Faegri & Iversen 1964; Pennington 1968; Tschudy & Scott 1969). In this way it is possible to discover how the vegetation changed with time (Moore & Bellamy 1974; Fig. 12.7; Wilkinson 1978). Sampling peat from known depths is possible with specially designed borers while the time scale can be checked by radiocarbon dating of wood fragments trapped in the peat at various levels.

(c) *Exploitation of peat deposits*

Domestic use

The value of dried peat as a fuel has been known for thousands of years in northern Europe. In the traditional technique for obtaining peat for this purpose the bog has first to be drained, usually a section at a time, by digging trenches. The *Sphagnum* and other vegetation is stripped off adjacent to one of the trenches and small sods of peat are cut out with a shaped spade and the placed on the top of the bog close by. After drying a little the blocks of peat are placed in

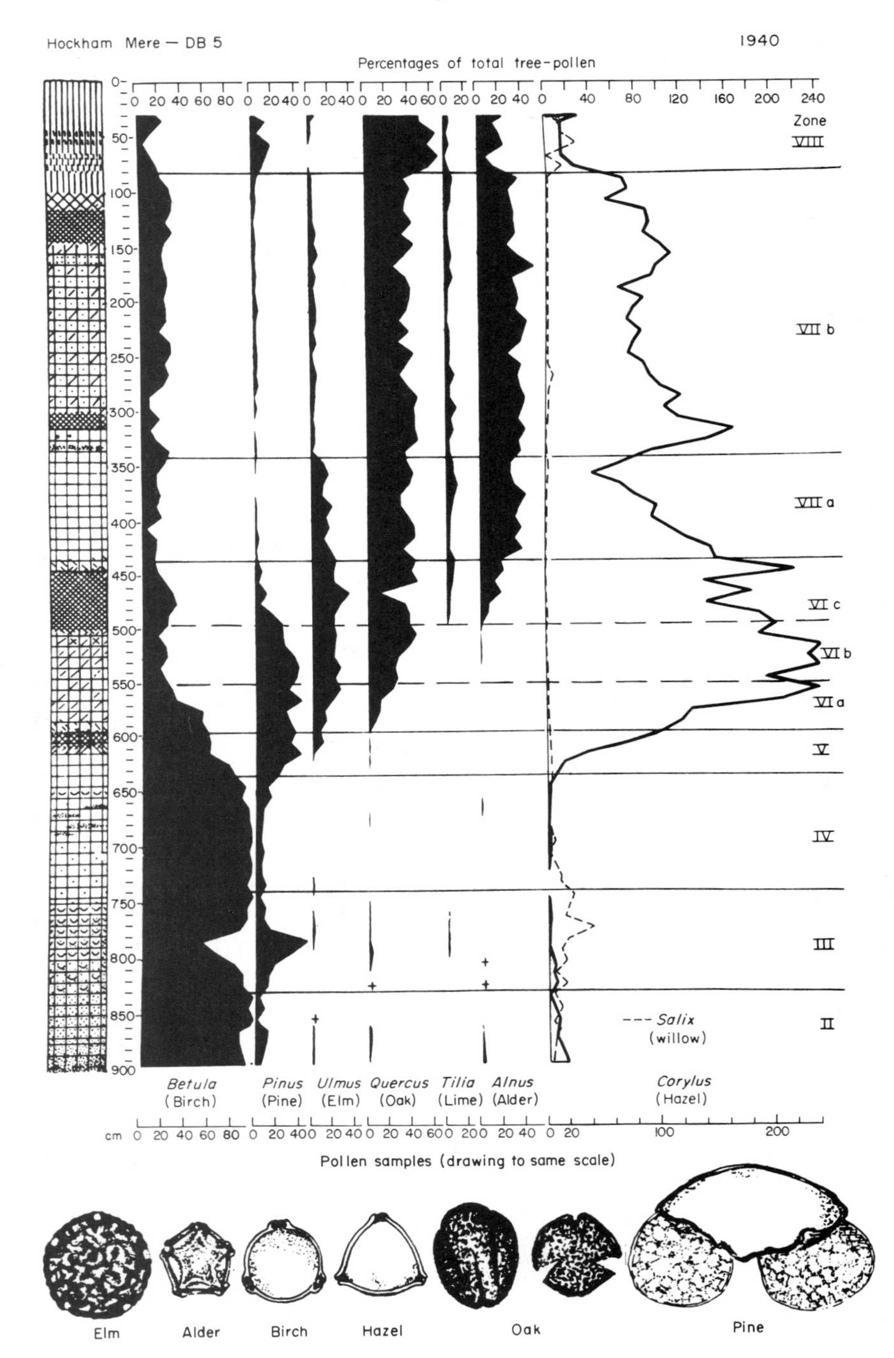
Hockham Mere — DB 5
1940
Percentages of total tree-pollen
Zone
VIII
VII b
VII a
VI c
VI b
VI a
V
IV
III
II
--- Salix (willow)
Betula (Birch)
Pinus (Pine)
Ulmus (Elm)
Quercus (Oak)
Tilia (Lime)
Alnus (Alder)
Corylus (Hazel)
cm
Pollen samples (drawing to same scale)
Elm
Alder
Birch
Hazel
Oak
Pine

windrows, turned, and collected using barrows and donkeys with paniers. The peat is then transported to village or town and stacked by the houses to be used when required. With moderate activity, it is said that a team of five men can lift and lay out 12,000 sods (85 m^3) of peat daily (Peckham 1874). The various stages mentioned above reduce the moisture content of the peat from about 95% in the undrained *Sphagnum* peat to about 30%. It is estimated that the annual production of such hand-won peat in Ireland amounts to about three and a half million tonnes a year (Miller 1957). Peat has a calorific value of about 3,300 cal/g, about half that of coal but much more than wood.

Commercial exploitation

In 1895 the prospectus for the Moss Litter and Peat Industries Company quoted figures to show that the use of peat in the U.K. for purposes other than fuel had doubled between 1890 and 1895 to 100,000 tonnes a year, most of which was used for stable bedding (Anon. 1895a). One problem was that the slightly decomposed 'red sphagnum peat' sold as stable litter contained fibre. The company developed a technique for removing the fibre from which they produced peat wool. This could be made into surgical dressings and 12,000 kg were said to have been sent for use by the French army during their expeditions to Madagascar between 1885 and 1896. Cloth could also be woven of peat wool and a dress of this material was apparently purchased by Her Royal Highness the Duchess of York. Other products made from peat at that time included ebony-like material, insulation for refrigerators, paper and peat briquettes (Anon. 1895b). In spite of all hopes the products did not become commercially successful.

Since the Second World War, Ireland has led the West in exploiting peat deposits, while the U.S.S.R. extracts the greatest amount of peat (100 million tonnes a year) (Anon. 1979a). The Irish Peat Development Authority (Bord na Mona) fosters research and development and now has 12 production areas covering 50,000 hectares (130,000 acres) of bog

Fig. 12.7 Vegetational changes in the flora of England with time as revealed by analysis of pollen grains preserved in *Sphagnum* mires. Pollen charts showing the amount of tree pollen as a percentage of the total. Hazel (*Corylus*) and willow (*Salix*) are treated separately as shrubs. Zones which cannot always be dated accurately establish the sequence. The elm decline in Zone VII is common to nearly all European pollen charts (Wilkinson 1978, adapted from Godwin 1975).

and employs 6,000 people who produce products worth more than 30 million pounds a year. These include 3 million tonnes of milled peat (55% moisture) which power seven electricity generating stations supplying about 20% of the nation's requirements (Fig. 12.8), a million tonnes of machine turf (35% moisture) which are small blocks of peat for domestic fuel, about half a million tonnes of peat briquettes (10% moisture) formed of compressed peat for steam and heat in industry, and one million cubic metres of light fibrous slightly decomposed moss peat for horticultural purposes. The latter is mainly exported to the U.K., Spain, the West Indies, the Middle East, Japan and Australia, earning 7 million pounds a year in foreign currency (Anon. 1979b, c).

Fig. 12.8 The Rhode Generating Station, County Offaly, Ireland, which is powered by milled *Sphagnum* peat delivered by the railway (on the left). 2000 tons a day are used (photo: Irish Electricity Supply Board).

This scale of operation demands the design of special machinery, especially since vehicles working on undrained bogs must not exceed pressures of 70g/cm² (1 lb/sq. in.) while even on drained bogs only 140g/cm² (2 lb/sq. in.) is the standard specification (Miller 1957). For machine peat a specially designed dredger excavates the peat, which is macerated and extruded, finally being cut into pieces by a cutting disc. For milled peat, the surface of the bog is cut into small pieces up to a depth of 1·5 cm by a milling machine. The surface is harrowed to accelerate drying, the peat scraped into ridges and collected by harvesting machines. The milled peat may be later dried further and compressed to form briquettes. Different machines are required to excavate moss peat without destroying its basic absorptive quality; mobile conveyor belts then collect and stack the crop for baling (Anon. 1979b, c).

In North America, the potential value of peat deposits in Minnesota was extolled as early as 1874 (Peckham 1874). However, it is only since the 'Energy Crisis' that serious attention has been given to exploiting this resource on a large scale. It is estimated that there are 60 billion tonnes of peat in the U.S.A. (excluding Alaska which has an equivalent amount) of which half is in Minnesota. In total this is equivalent to 240 billion barrels of oil, and for this reason plans are being put into operation to exploit the peat deposits in eastern North Carolina and Minnesota, mainly to produce synthetic natural gas and to fuel power stations (Anon. 1975a, 1978). In Minnesota there are some 1000 km^2 (400 square miles) of peat deposits with an average depth of 2 m (6 m in places) and there are more than 50 bogs whose area is in excess of 40,000 hectares (100,000 acres). This dwarfs the situation in Ireland where not a single bog exceeds 40,000 hectares. Other countries, including Russia, Finland and Canada, are harvesting peat in commercial quantities. The value of renewable resources, such as that based on *Sphagnum*, is now being recognized since it is clear that alternatives to oil are going to be needed in the near future.

3 CONSERVATION

3.1 PEAT MOSSES

The exploitation of peat on a scale envisaged in many countries is of great concern to ecologists as it involves destruction of both living *Sphagnum* and all the associated flowering plants which form the vegetation on peat deposits. The electric power generated may be used for industrial purposes resulting in emissions of toxic gases such as SO_2 that may also result in the death of unexploited *Sphagnum*. Indeed the disappearance of all but the resistant *Sphagnum recurvum* from the

Pennine upland mires in England is thought to be due to this (Ferguson *et al.* 1978). On a local scale trampling by quite small numbers of people can radically alter mire plant communities. Even students on field trips or doing research can inadvertently cause damage. However such problems have been solved successfully in many places by the use of duckboarding and raised walkways (Moore & Bellamy 1974).

The destruction of a large percentage of the world's very productive coastal wetlands took place through drainage, building and pollution before the importance of these areas was fully realized. It is to be hoped that energy-hungry man of the present century will not make a similar mistake with peatlands.

3.2 OTHER MOSSES

(a) *Habitat destruction*

Habitat destruction is probably the most important factor which has lead to the disappearance of many moss species from places where they were once common. Clear felling of woodlands is particularly harmful for species which depend on the shade and high humidity found beneath trees. Even if replanting occurs the bryophyte flora is affected. For example the recent felling of the bryophyte rich oakwoods at Blackstone Bridge, County Kerry, Ireland resulted in the death within two years of all but the most tolerant species due to scorching and drought induced by the summer sun (Ratcliffe 1968). In temperate regions, even if clear felling is not practiced, underplanting by exotic coniferous trees can be harmful since both the pH and the character of the bark are foreign to the native corticolous mosses.

The burning of moorland to improve sheep pasture leads to changes in the bryophyte communities as does overgrazing by domesticated stock. Damming streams to conserve water can affect the aquatic mosses but drainage improvement is worse as it often results in floods during periods of heavy rain which can scour off bryophyte growth and redistribute boulders on which mosses grow. Forty years after a dam collapsed in the Conway Valley, Wales, the mosses had only reached the initial stages of colonization in the Porth Llwyd ravine (Ratcliffe 1968).

(b) *Collecting*

The demand for mosses by horticulturists and florists for packing material and wreaths seems to be growing. The suppliers of this commodity may strip large masses of carpet moss from rocks, walls and tree bases leaving conspicuous bare patches as may be seen in some parts of Wales and the Lake District in Britain. Scientific supply companies should also be included among the commercial moss gatherers. Generally, only common genera are collected in significant amounts, e.g. *Sphagnum*, *Andreaea*, *Polytrichum*, *Mnium* and *Funaria*. However very large quantities of material are needed for the 1st and 2nd year biology courses at the larger North American universities where student numbers may reach several hundred per class. Furthermore, material is normally required with mature sex organs or capsules so that the reproductive potential is likely to be affected in the moss communities from which collections are taken.

Bryologists or their students are not free from blame either as they tend to be avid collectors of rare species which often end up in poorly curated private herbaria which are discarded on the death of the owner. The paucity of certain species around some field stations visited annually by student groups attests to long term impact of such activities. Professional taxonomists are not above reproach, occasionally collecting large quantities to prepare exsiccati and leaving unsightly damage to the trees on which the epiphytes grow. Incensed local residents in Costa Rica recently received a substanital out-of-court settlement from a group of over-enthusiastic botanists on an international field trip (Nash & Dibben 1979). Such activities can reduce if not exterminate rare species in certain localities. The need for collecting can be fully justified by taxonomists engaged in specific research projects but those not involved in this branch of bryology should be armed with a camera rather than collecting bags.

(c) *Pollution*

Another factor mentioned in previous chapters is the increasing levels of air and water pollution on a local and global scale. Sewage and chemical waste have the greatest effect on aquatic mosses while sulphur dioxide is generally the most harmful component of air pollution for terrestrial mosses. The latter may cause the death of mosses close to an emission source (see Chapter 10) and is probably the cause of the observed reduction in abundance of sensitive epiphytic species of *Ulota* and *Orthotrichum* in Britain since the advent of the industrial revolution (Ratcliffe 1968). It is depressing that the plans which proposed that a fossil-fuelled power station (without stack scrubbers) be sited at Plymouth, England, a few miles to windward of Wistmans wood were only quashed by a turndown in the economy (Anon. 1972). The wood whose primaeval beauty is renowned and depends largely on the rich epiphytic flora of mosses and lichens, is of recognized scientific interest and visited by hundreds of people annually (Proctor *et al.* 1980).

The many man-induced changes cited above which are deleterious to the continued existence of rich moss floras will no doubt proceed. However it is hoped that a greater understanding of the biology of mosses including their interactions with other plants and animals will lead to a more thoughtful approach to planning as well as to conservation programmes aimed at preserving what remains of threatened habitats whose bryophyte flora is of scientific or aesthetic value.

4 REFERENCES

ANDO H. (1971) Les jardins de mousses au Japon. Pl. *Montagne Bull. Soc. Amat. Jard. Alpins.* **5,** 290–4.
ANDO H. (1980) Bibliography of the use of bryophytes. *Hikobia,* **8,** 424–48.
ANON. (1822) *The Moss House.* William Darton, London.
ANON. (1867) *The Moss Gatherers.* Society for promoting Christian knowledge, London.
ANON. (1895a) *Prospectus for Moss Litter and Peat Industries Co., Ltd.* Grieve and Pass, London.
ANON. (1895b) A new industry. *Daily Telegraph,* London, 26 July 1895.
ANON. (1968) Keen bidding for right to gather moss. *Guardian,* London, 19 February, 1968.
ANON. (1972) Sulphur dioxide emissions reach highest level ever. *Br. Lichen. Soc. Bull.* **31,** 1–2.
ANON. (1975a) Energy: plan to use peat as fuel stirs concern in Minnesota. *Science* (*Wash. D.C.*), **190,** 1066–70.
ANON. (1975b) Moss underfoot—cool, green, wild. *Sunset Magazine,* S. Calif. (Hawaii Ed.) July 1975, 136–7.
ANON. (1978) Peat for fuel: development pushed by big corporation farm in Carolina. *Science* (*Wash. D.C.*), **199,** 33–4.
ANON. (1979a) Peat for fuel. *Time* (*N.Y.*), **114, (23),** 38–9.
ANON. (1979b) Mimeographed information provided by Irish Peat Development Authority.
ANON. (1979c) 32nd Annual Report Bord na Mona. Dublin.
ANONYMOSS (1979) Mobile homes for mosses. *Bull. Br. Bryol. Soc.* **33,** 6.
BAGNALL J.E. (1889) *Handbook of Mosses.* Swan Sonnenschein, London.
BLAND J.H. (1971) *The forests of Lilliput: the Realm of Mosses and Lichens.* Prentice Hall, Englewood Cliffs, New Jersey.
BOND T.E.T. (1976) *Polytrichum* spp. in an apple orchard on herbicide-treated soil. *Proc. Bristol Naturalists' Soc.* **35,** 69–72.
BRITTON E.G. (1902) *Climacium dendroideum* for millinery. *Bryologist,* **5,** 98.
CLARKE C.H. (1902) Bryological Millinery. *Bryologist,* **5,** 77–8.
CLYMO R.S. (1970) The growth of *Sphagnum*: methods of measurement. *J. Ecol.* **58,** 13–49.
CLYMO R.S. (1978) A model of peat bog growth. *Ecol. Stud.* **27,** 187–223.

COOK C.D.K., GUT R.J., RIX E.M., SCHNELLER J. & SEITZ M. (1974) *Water Plants of the World. A manual for the identification of the genera of fresh water macrophytes.* W. Junk, The Hague.

CRUM H. (1972) The geographic origins of mosses of North America's eastern deciduous forest. *J. Hattori Bot. Lab.* **35,** 269–98.

CRUM H. (1973) Mosses of the Great Lakes Forest. *Contrib. Univ. Mich. Herb.* **10,** 1–404.

DICKSON J.H. (1973) *Bryophytes of the Pleistocene.* Cambridge University Press.

DIMBLEBY G. (1967) *Plants & Archaeology.* Granada Publishing, London.

FAEGRI K. & IVERSEN J. (1964) Textbook of pollen analysis. Munksgaad, Copenhagen.

FERGUSON P., LEE J.A. & BELL J.N.B. (1978) Effects of sulphur pollutants on the growth of *Sphagnum* species. *Environ. Pollut.* **16,** 151–62.

FLOWERS S. (1957) Ethnobryology of the Gosuite Indians of Utah. *Bryologist,* **60,** 11–14.

FRYE T.C. (1920) Notes on useful and harmful mosses *Bryologist,* **23,** 71.

GLOB P.V. (1971) *The Bog People.* Granada Publishing, London.

GODWIN H. (1975) *History of the British Flora.* 2nd ed., pp. 44, 47. Cambridge University Press.

HOTSON J.W. (1921a) *Sphagnum* used as a surgical dressing in Germany during the World War. *Bryologist,* **24,** 74–8.

HOTSON J.W. (1921b) *Sphagnum* used as a surgical dressing in Germany during the World War. *Bryologist,* **24,** 89–96.

INOUE H. (1972) The special moss exhibition displayed at the National Science Museum, Tokyo, 1971. *Bryologist,* **75,** 1–6.

ISHIKAWA I. (1974) Bryophyta in Japanese gardens (2). *Hikobia,* **7,** 65–78.

IWATSUKI Z. & KODAMA T. (1961) Mosses in Japanese gardens. *Econ. Bot.* **15,** 264–9.

MILLER H.M.S. (1957) Peat. In *A View of Ireland,* eds., Meenan J. & Webb D.A. pp. 125–48. British Association for the Advancement of Science, Dublin.

MOORE P.D. & BELLAMY D.J. (1974) *Peatlands.* Springer Verlag, New York.

NASH T.H. & DIBBEN M.J. (1979) On the ethics of group lichen collecting. *Int. Lichenol. Newsletter,* **12(1),** 1–3.

NICHOLS G.E. (1918a) War work for Bryologists. *Bryologist,* **21,** 53–6.

NICHOLS G.E. (1918b) The American Red Cross wants information regarding the supplies of surgical *Sphagnum. Bryologist,* **21,** 81–3.

PARK M. (1799) *Travels in the Interior Districts of Africa.* Bulmer, London.

PECKHAM S.F. (1874) Peat for Domestic Fuel. The Geological and Natural History survey of Minnesota. University of Minnesota.

PENNINGTON W. (1969) *The History of British Vegetation.* English Univ. Press, London.

PROCTOR M.C.F., SPOONER G.M. & SPOONER M.F. (1980) Changes in Wistman's wood, Dartmoor: photographic and other evidence. *Rep. Trans. Devon Assoc. Advmt. Sci,* **112,** 43–79.

RATCLIFFE D.A. (1968) An ecological account of atlantic bryophytes in the British Isles. *New Phytol.* **67,** 364–439.

RICHARDSON D.H.S. (1975) *The Vanishing Lichens,* David & Charles, Newton Abbot.

ROBINSON D.W. & O'KENNEDY N.D. (1978) The effect of overall herbicide systems of soil management on the growth and yield of apple trees 'Golden Delicious'. *Scientia Hortic.* **9,** 127–36.

SEAWARD M.R.D. (1976) *The Vindolanda Environment.* Barcombe Publications, Haltwhistle.

SEAWARD M.R.D. & WILLIAMS D. (1976) An interpretation of mosses found in recent archaeological excavations. *J. Archaeol. Sci.* **3,** 173–7.

SEAWARD M.R.D. (1979) Lower plants and the urban landscape. *Urban Ecol.* **4,** 217–25.

SHIU-YING H. (1945) Medicinal plants of Chengtu herb shops. *J. West China Border Res. Soc. B.* **15,** 95–176.

STUBBS J. (1973) Moss Control. *New Sci.* **57,** 739.

THIERET J.W. (1956) Bryophytes as economic plants. *Econ. Bot.* **10,** 75–91.

TRIPP F.E. (1874) *British Mosses, their Homes Aspects Structure and Uses.* vol. 1. p. 23, George Ball, London.

TSCHUDY R.H. & SCOTT R.A. (1969) *Aspects of Palynology.* Wiley Interscience.

WALKER D. & WEST R.G. (1970) *Studies in the Vegetational history of the British Isles.* Cambridge University Press.

WATLING R. & SEAWARD M.R.D. (1976) Some observations on puff-balls from British Archaeological sites. *J. Archeol. Sci.* **3,** 165–72.

WHITE G. (1832) *The Natural History of Selbourne,* p. 143. The London Folio Society, London.

WILKINSON G. (1978) *Epitaph for the Elm.* Hutchinson, London.

APPENDIX

The orders of true mosses (Bryopsida) and the characters typical of each order are shown in the table on the following pages. Also included are examples of genera which are placed in the orders.

Order	Shoots	Leaves	Leaf cells	Seta	Capsule	Peristome	Calyptra	
Subclass 1. POLYTRICHIDEAE								
Tetraphidales	Acrocarpous small (2 cm)	Egg-shaped to lance-shaped; nerve when present without lamellae on surface	Round to hexagonal, smooth in a single layer	Long	Cylindrical, erect, without annulus	4 teeth	Conical and symmetrical with 2 or more slits	*Tetraphis*
Polytrichales	Acrocarpous, frequently large (10 cm) with tough stems showing internal anatomical differentiation	Lance-shaped, nerve with lamellae on ventral surface often with sheathing base	Lamellae in central region, marginal cells square to roundish	Long	Spherical to cylindrical, round or 2–6 angled in cross section	Columella expanded at top into membranous epiphram which is joined to the tips of the 32 or 64 teeth	Smooth or hairy	*Atrichum* *Dawsonia* *Polytrichum*
Subclass 2. BUXBAUMIIDEAE								
Buxbaumiales	Acrocarpous, small to minute (2 cm) plants	Often not well developed, those around sex organs have cilia on margin	Rounded to hexagonal	Short or long often rough surface	Egg-shaped, inclined	Single or double; outer peristome 1–4 rows of thread-like teeth, faintly transverse barred. Inner peristome a folded membrane	Very small	*Buxbaumia*

SUBCLASS 3. EUBRYIDEAE—orders with peristomes which are normally absent or single

Archidiales	Small (2 cm), stems arising from below the leaves surrounding the sex organs	Lance-shaped with sheathing base and a nerve	Thick walled, rectangular to hexagonal	Very short	Roundish without a lid	Absent and no columella	Small and soon shrivels	*Archidium*
Dicranales	Large (10 cm), having apical cell with 3 outing faces	Needle-shaped to broadly egg-shaped with a nerve	Upper cells square to rectangular; basal cells square to elongate sometimes with pores	Long	Egg-shaped to cylindrical, frequently curved usually erect	Single. 16 teeth usually divided into two segments with transverse bars and vertical striae between the bars	Hooded or conical and symmetrical with two or more slits	*Ceratodon* *Dicranum* *Dicranoweisia* *Leucobryum*
Fissidentales	Small or moderate (5 cm) apical cell with 2 cutting faces	Leaves in 2 rows with a nerve oblong to lance shaped	Cells hexagonal, one or two layers, smooth or with papillae Sometimes a border of narrow cells	Long or short	Ellipsoid, erect or inclined	Single. 16 teeth usually divided into 2 segments with transverse bars and vertical striae between the bars	Hooded or conical and symmetrical with 2 or more slits	*Fissidens*
Pottiales	Acrocarpous, usually small (3 cm)	Egg-shaped to linear with pointed to rounded tip and nerve	Upper cells hexagonal or square often with papillae, lower cells rectangular	Long or short	Erect or inclined with or without a lid, egg-shaped to cylindrical	Single, 16 or 32 teeth often spirally twisted and divided into thread-like segments	Hooded	*Acaulon* *Barbula* *Desmadoton* *Pottia* *Tortella* *Tortula* ***Weissia***

Order	Shoots	Leaves	Leaf cells	Seta	Capsule	Peristome	Calyptra	
Grimmiales	Acrocarpous usually on rock large (10 cm)	Egg-shaped to linear often with long hair points and nerve	Upper cells square or round with thick walls; lower cells longer and thinner walls	Long or short	Sperical to cylindrical	Single, 16 teeth usually divided into 2 segments with transverse bars, no vertical striae	Hooded or conical and symmetrical with 2 or more slits	*Grimmia* *Rhacomitrium*
Encalyptales	Acrocarpous moderate size (5 cm)	Lance-shaped to strap shaped, pointed with a nerve	Upper cells hexagonal to square very papillose. Lower cells rectangular, marginal cells narrow.	Long	Erect and cylindrical	Variable, single or absent sometimes double	Huge and covering whole capsule often fringed or notched at base	*Encalypta*
Schistostegales	Plants arising from a persistent thalloid protonema with light refracting properties, small (2 cm)	In two ranks egg-shaped to lance-shaped without a nerve	Rhomboidal	Thin long	Spherical erect	Absent		*Schistostega*
SUBCLASS 4. EUBRYIDEAE—orders with peristomes which are normally double.								
Funariales	Acrocarpous soil usually small (3 cm) except Splachnaceae which are large (6 cm)	Egg-shaped to lance-shaped with thin nerve	Large thin walled hexagonal to rhomboid	Long or short	Erect or inclined may be asymmetrical round to pear shaped with a distinct neck	Variable, single or double of 16 teeth each or absent. Teeth may be attached to central disc	Large, hooded or symmetrical with slits	*Aplodon* *Ephemerum* *Funaria* *Physcomitrella* *Physcomitrium* *Splachnum* *Tayloria* *Tetraplodon*

Bryales	Acrocarpous moderate to large (10 cm)	Egg-shaped to lance-shaped often with a border and with a nerve	Hexagonal to linear	Long	Spherical to cylindrical, inclined to drooping	Double each of 16 teeth, the outer teeth with horizontal bars; inner peristome a folded membrane which continues as cilia	Hooded, small and soon lost	*Aulacomnium* *Bryum* *Leptobryum* *Mielichhoferia* *Mittenia* *Mnium* *Philonotis* *Plagiobryum* *Plagiomnium* *Pohlia* *Rhodobryum* *Timmia* *Orthotrichum*
Orthotrichales	Acrocarpous often on trees moderate (5 cm)	Egg-shaped to lance-shaped with nerve	Upper cells rounded or hexagonal, lower cells elongate	Short to long	Erect ellipsoid tapering into seta	Double, 16 outer teeth often united in pairs with papillae or striae. Inner teeth thin often missing	Large bell shaped often with coarse erect hairs	*Ulota*
Isobryales	Pleurocarpous primary stems creeping but secondary stems may be erect. Large (5 cm)	3 or more rows on stem. Egg-shaped to lance-shaped with pointed tip. Nerve present or absent	Square to elongate sometimes with papillae	Long or short	Round to cylindrical	Double, 16 outer teeth with papillae or horizontally striate. Inner teeth thread-like may be united into lattice	Hooded or symmetrical with slits	*Climacium* *Fontinalis* *Neckera*

Order	Shoots	Leaves	Leaf cells	Seta	Capsule	Peristome	Calyptra	
Hookeriales	Pleurocarpous moderate (5 cm)	Often flattened into one plane, sometimes with border. Egg-shaped to lance-shaped with pointed tip. Single or double nerve	Rhomboidal, elongate near margin	Long	Erect to inclined	Double, outer teeth with median furrow inner teeth without cilia	Conical with a fringe at the base	*Hookeria*
Thuidiales	Pleurocarpous plants minute to robust (10 cm)	Egg-shaped usually pointed with nerve, sometimes double or absent	Rounded to linear usually with papillae	Long	Erect or inclined egg-shaped to cylindrical sometimes curved	Double, inner with cilia	Hooded	*Anomodon* *Thuidium*
Hypnobryales	Pleurocarpous slender to robust (20 cm)	Lance-shaped to rounded with a single, or double nerve or without one	Rhomboidal to elongate	Long	Inclined egg-shaped to cylindrical	Double, outer teeth with papillae, inner with or without cilia	Hooded	*Brachythecium* *Cratoneuron* *Drepanocladus* *Eurhychium* *Hylocomium* *Hypnum* *Pleurozium* *Pseudosclero-podium* *Rhitidiadelphus*

Data abstracted from Smith 1978.

GLOSSARY

Acclimation. The ability of a plant to alter its physiology so that optimum rates of photosynthesis occur under the conditions of light or temperature that prevail at each season of the year.

Acrocarpous (Gk. *acros*, topmost + *karpos*, fruit). Mosses which generally form erect tufts that develop antheridia and archegonia, and subsequently capsules, at the apices of the main branches.

Allele (Gk. *allelon*, of one another). One of the two alternative states of a gene on a chromosome. Alleles are separated from one another during meiosis.

Annulus (L. a ring). A ring of specialised cells involved in throwing off the lid of the capsule.

Apophysis (Gk. *apo*, separate + *physis*, growth). The more or less swollen portion of a capsule, below the spore sac, where photosynthesis occurs most actively.

Apospory (Gk. *apos*, separate + *spora*, a seed). The production of a gametophyte directly from a sporophyte not involving the formation of spores.

Apogamy (Gk. *apos*, separate + *gamete*, a wife). The production of a sporophyte from a gametophyte directly without the involvement of fusion between egg and sperm.

Antheridium (Gk. *anthos*, flower + *idium*, diminutive ending). Sperm-containing sex organ in mosses.

Archegonium (Gk. *archegonas*, founder of the race + *idium*, diminutive ending). The female sex organ of mosses which contains and protects the egg.

Autoradiography (Gk. *autos*, self + L. *radiolus*, a ray + Gk. *graphe*, a painting). A photographic print made on sensitive film by radiation from a radioactive substance (usually administered to a plant in the laboratory).

Calyptra (Gk. *kaliptra*, covering for the head). Hood or cap partly or entirely enveloping the capsule and formed from the expanded archegonial wall.

Capsule. Spore containing structure of mosses usually differentiated into a lower sterile part, the apophysis, and upper spore producing part.

Caulonema (Gk. *caulos*, a stem + *nema*, a hair). The type of protonemal filament, formed about a week after moss spore germination, which has few chloroplasts and cell walls which are oblique. Buds develop on the filaments and grow into leafy shoots.

Chloronema (Gk. *chloros*, green + *nema*, a hair). The type of protonemal filament first formed on germination of a moss spore. It has many chloroplasts and cell walls which are perpendicular to the long axis of the filament.

Chromosomes (Gk. *chromos*, colour + *some*, body). The bodies in the cell nucleus which appear during division and which bear the genes that determine hereditary characters.

Columella. Central mass or column of tissue in the capsule which is surrounded by the sac of spores.

Compensation period. The length of time required by a plant for photosynthesis, in order that night-time losses of carbon as a result of respiration are replaced.

Compensation point (light). The intensity of light which results in such a photosynthetic rate that carbon dioxide fixation by this process exactly balances carbon dioxide output due to respiration.

Cytogenetics (Gk. *kytos*, container + *genos*, offspring). The field of investigation concerned with studies of chromosomes.

Dioecious (Gk. *di*, two + *oikos*, house). Having male (antheridia) and female (archegonia) sex organs borne on separate plants.

Diploid (Gk. *diploos*, double + *oides*, like). Having a double set of chromosomes as is frequently found in sporophytes.

Ectohydric (Gk. *ectos*, outside + *hydro*, water). Possession of little internal conducting tissue so that water is absorbed through stem and leaf surfaces rather than being transported from the base of the plant.

Endohydric (Gk. *endon*, within + *hydro*, water). Possession of a well-developed conducting system within the stem which enables water movement from base to apex of the plant.

Epiphragm (Gk. *epi*, upon + *phrasso*, a hedge). A disc like plate of tissue, formed from the top of the columella, at the mouth of the capsule of some mosses. It is attached to the ends of the peristome teeth.

Epiphyte (Gk. *epi*, upon + *phyton*, plant). A plant which grows upon another plant merely using the latter as a means of support.

Exine. The outer layer of the wall of moss spores.

Foot. The lowermost portion of the capsule stalk which is embedded in the gametophore and acts as an anchoring and absorptive organ.

Gametophore (Gk. *gamete*, a wife + *phoros*, bearing). Upright green leafy stems which develop from buds on the protonema and eventually form sex organs.

Gametophyte (Gk. *gamete*, a wife + *phyton*, a plant). The sex-organ-producing, generation of mosses consisting of rhizoids, gametophores, antheridia and/or archegonia.

Haploid (Gk. *haploos*, single + *oides*, like). Having a single set of chromosomes as in most gametophytes.

Hybrids. Offspring of two different varieties or species.

Hydroids. The water conducting cells in the hydrom.

Hydrom (Gk. *hydros*, water). The water conducting tissue in moss stems and rhizomes.

Hygroscopic (Gk. *hygros*, wet + *scopos*, observation). The tendency to absorb water from moist air.

Gemmae (L. a bud). A vegetative propagule which becomes detached and can form a new plant.

Intine. The inner layer of the wall of moss spores that is made up of cellulose.

Labyrinth. A complicated series of infoldings formed by the growth of cell wall material into the cell so increasing the area of the cell membrane for better absorption or transfer.

Leptoids. The food conducting cells within the leptom.

Leptom. The food conducting tissue in moss stems and rhizomes.

m-chromosomes. Minute chromosomes usually 1/10 or less the size of other chromosomes.

Meiosis (Gk. *meoun*, to make smaller). A process which in mosses takes place during spore formation. It ensures that the spores (the first cells of the gametophyte generation) have half the compliment of chromosomes as the cells of the capsule (part of the sporophyte). See also haploid and diploid.

Monoecious (Gk. *monos*, single + *oikos*, house). Having male and female sex organs (antheridia and archegonia) on the same plant.

Nomarsky optics. A microscopic procedure by which unstained components of cells can be distinguished by their differential interference patterns.

Oogenesis (Gk. *oion*, egg + *genos*, offspring). The process in which eggs are formed.

Operculum (L. *operculum*, lid). The lid of the moss capsule which, when shed, often reveals a peristome.

Paraphyses (Gk. *para*, beside + *physis*, growth). Sterile filaments growing among the antheridia or archegonia which help to protect the sex organs and prevent them drying out.

Perigonial leaves (Gk. *peri*, around + *gonos*, a child). The modified leaves which surround antheridia at the apex of the gametophore.

Perine. The outermost layer of the spore wall of mosses which is often sculptured and ornamented and which may be composed of the remnants of the cells in which the spores formed.

Peristome (Gk. *peri*, around + *stome*, a mouth). A ring, often double, of teeth involved in spore dispersal that is found around the opening of a moss capsule.

Pleurocarpous (Gk. *pleuros*, side of body + *karpos*, fruit). Mosses which have a tendency for dorsi-ventral orientation (carpet forming) and whose archegonia and subsequently capsules form on short lateral branches.

Poikilohydric (Gk. *poikilos*, various + *hydro*, water). Noun; a plant whose water content varies with that of the surrounding environment.

Poikilohydrous. Adjective; see poikilohydric.

Polyploid (Gk. *polys*, many + *ploos*, fold). A cell with several complete sets of chromosomes.

Protonema (Gk. *protos*, first + *nema*, thread). Green, usually filamentous, thread produced by a germinating spore or regenerating part.

Pseudopodium (Gk. *pseudes*, false + *podion*, foot). A prolongation of gametophore apex which elevates the capsules in place of the seta, which in some mosses remains short.

Rhizoid (Gk. *rhiza*, root). Hair-like structure which attaches moss plants to the substratum.

Seta (L. *seta*, bristle). The stalk on which most moss capsules are elevated.

Sporogenesis (Gk. *spora*, seed + *genos*, offspring). The process in which spores are formed.

Sporophyte (Gk. *spora*, a seed + *phyton*, a plant). The spore producing diploid generation consisting of a foot, seta and capsule.

Stalk. see seta.

Succession. The slow orderly progression of change in the communities of plants that occurs during the development of vegetation, in an area, from initial colonization.

Tuber (L. swelling). Vegetative propagule on rhizoids or underground stem which can grow into a new plant.

Water potential. The chemical activity of water in a particular system as compared with that of pure water under standard conditions. By measuring the water potential of cells, it is possible to assess the degree of water stress to which the plant is subjected. The more negative the water potential, expressed in bars, the greater the stress.

INDEX